HELM FIELD GUIDES

Birds of
CHILE

Daniel Martínez Piña and
Gonzalo González Cifuentes

HELM
LONDON · OXFORD · NEW YORK · NEW DELHI · SYDNEY

HELM
Bloomsbury Publishing Plc
50 Bedford Square, London, WC1B 3DP, UK

BLOOMSBURY, HELM and the Helm logo are trademarks of Bloomsbury Publishing Plc

First published in Great Britain 2021

Copyright © Daniel E. Martínez Piña and Gonzalo E. González Cifuentes 2021
Illustrations © Daniel E. Martínez Piña 2021

The authors have asserted their right under the Copyright, Designs and Patents Act, 1988, to be identified as Authors of this work.

All rights reserved. No part of this publication may be reproduced or transmitted in any form or by any means, electronic or mechanical, including photocopying, recording, or any information storage or retrieval system, without prior permission in writing from the publishers.

Bloomsbury Publishing Plc does not have any control over, or responsibility for, any third-party websites referred to in this book. All internet addresses given in this book were correct at the time of going to press. The authors and publisher regret any inconvenience caused if addresses have changed or sites have ceased to exist, but can accept no responsibility for any such changes.

A catalogue record for this book is available from the British Library.

Library of Congress Cataloguing-in-Publication data has been applied for.

ISBN: HB: 978-1-4729-8742-6
PB: 978-1-4729-7000-8
ePub: 978-1-4729-6999-6

2 4 6 8 10 9 7 5 3 1

Design by Susan McIntyre
Printed and bound in India by Replika Press Pvt. Ltd.

To find out more about our authors and books visit www.bloomsbury.com and sign up for our newsletters.

CONTENTS

	Plate	Page
ACKNOWLEDGEMENTS		7
FOREWORD		9
INTRODUCTION		10
Species included		10
Taxonomy		11
Structure of the book		11
Administrative regions of Chile		15
Biogeographic zones of Chile		16
Bird topography		17
SPECIES ACCOUNTS AND PLATES		
Penguins SPHENISCIDAE	1–2	18–20
Albatrosses DIOMEDEIDAE	3–8	22–31
Petrels and shearwaters PROCELLARIIDAE	4, 9–12, 14	24, 32–38, 42
Northern storm petrels HYDROBATIDAE	13	40
Southern storm petrels OCEANITIDAE	13–14	40–42
Cormorants and shags PHALACROCORACIDAE	15	44
Pelicans PELECANIDAE	16	46
Frigatebirds FREGATIDAE	16	46
Boobies SULIDAE	17	48
Tropicbirds PHAETHONTIDAE	17	48
Jaegers and skuas STERCORARIIDAE	18	50
Sheathbills CHIONIDAE	19	52
Gulls and terns LARIDAE	19–24	52–61
Skimmers RYNCHOPIDAE	21	56
Ducks, geese and swans ANATIDAE	25–29	62–70
Grebes PODICIPEDIDAE	30	72
Rails, crakes and gallinules RALLIDAE	31–32	74–76
Flamingos PHOENICOPTERIDAE	33	78
Spoonbills THRESKIORNITHIDAE	33	78
Herons and bitterns ARDEIDAE	34–35	80–82
Ibises THRESKIORNITHIDAE	35	82
Oystercatchers HAEMATOPODIDAE	36	84
Stilts and avocets RECURVIROSTRIDAE	36	84
Sandpipers and allies SCOLOPACIDAE	37–39, 42–43	86–90, 96–98
Plovers and lapwings CHARADRIIDAE	40–42	92–96
Magellanic Plover PLUVIANELLIDAE	41	94
Painted-snipes ROSTRATULIDAE	42	96
Thick-knees BURHINIDAE	43	98
Seedsnipes THINOCORIDAE	43	98
New World quails ODONTOPHORIDAE	43	98
Pheasants PHASIANIDAE	43	98

Rheas RHEIDAE	44	100
Tinamous TINAMIDAE	44	100
New World vultures CATHARTIDAE	45	102
Hawks, harriers and kites ACCIPITRIDAE	46–50	104–112
Osprey PANDIONIDAE	50	112
Falcons and caracaras FALCONIDAE	51–52	114–116
Raptors in flight	53–54	118–119
Pigeons and doves COLUMBIDAE	55–56	120–122
New World parrots PSITTACIDAE	57	124
Barn owls TYTONIDAE	58	126
Owls STRIGIDAE	58–59	126–128
Nightjars CAPRIMULGIDAE	59	128
Woodpeckers PICIDAE	60	130
Kingfishers ALCEDINIDAE	60	130
Swallows and martins HIRUNDINIDAE	61–62	132–134
Swifts APODIDAE	62	134
Hummingbirds TROCHILIDAE	62–63	134–136
Tapaculos RHINOCRYPTIDAE	64–65	138–140
Ovenbirds FURNARIIDAE	66–71	142–152
Tyrant flycatchers TYRANNIDAE	70–76	150–162
Thrushes TURDIDAE	77	164
Mockingbirds MIMIDAE	77	164
Cotingas COTINGIDAE	78	166
New World blackbirds ICTERIDAE	78–79	166–168
Cuckoos CUCULIDAE	79	168
Pipits MOTACILLIDAE	80	170
Wrens TROGLODYTIDAE	80	170
New World warblers PARULIDAE	81	172
Tanagers, finches and seedeaters THRAUPIDAE	81–86	172–182
New world sparrows PASSERELLIDAE	84	178
Old World sparrows PASSERIDAE	84	178
Siskins FRINGILLIDAE	87	184
Eggs of Chilean birds	88–89	186–187

EXTREME VAGRANTS   188

Appendix I: Endemic bird species in Chile   192
Appendix II: Endemic bird subspecies in Chile   193
Appendix III: Threatened bird species in Chile   195

CHECKLIST OF THE BIRDS OF CHILE   198

INDEX   213
QUICK INDEX TO THE MAIN GROUPS OF BIRDS   224

ACKNOWLEDGEMENTS

Many hours spent in front of a desk have inevitably separated me from my closest loved ones, and who must first receive my deepest thanks. I am, of course, referring to Constanza and our children Leonora and Mateo. My parents, brothers and close friends are also incredibly important to me, and are always very attentive to my work – thank you.

Once again, I must dedicate special thanks to Gonzalo González – he is both a friend, critic and fellow worker at all times.

In the original Spanish edition of this work, *Aves de Chile* (2017), you can find the names of a large number of direct collaborators who helped with the development of this work. I reiterate my thanks here for their investment and collaboration.

In this new English version, it only remains to thank Bloomsbury Publishing for their interest and support in the production of this book.

Daniel Martínez Piña

I would like to thank my family for their daily support, which is, *per se*, exhausting. Patricia Tagle, Sebastián González and Elsa Cuevas P. – thank you very much! Although I don't tell you often, I am incredibly grateful.

Thanks also to my friends and companions on birding trips around the country: Alexander Baus, Flavio Camus, Pamela Cartagena, and special thanks to Jorge Toledo, among others. Thank you for your time, your invaluable company and, above all, your friendship.

Many people – some of whom I only know via mail – have been incredibly helpful with the small details, which, as we know, are what help to define the whole. I would like to thank Mauro Bianchini who contributed articles and photographs of great use; Koky Castañeda for his images of White-winged Diuca-Finch; Marcelo Flores for the information and image of an Andean Flicker in the Tarapacá Region; Douglas Hardy for his images of a nesting White-winged Diuca-Finch; Steve N. G. Howell for his thoughtful responses to my various enquiries; Gabriela Ibarguchi for her kind response regarding seedsnipes; Victor Raimilla for his support with audio and images; Daniel González A. for his support of the project, which translated into supplying images, quotes, papers and more; Jorge Toledo for his reviews, comments, photographs and more; Claudia Godoy R. for the background information on King Penguins; Rodrigo Tapia for his images of terns; Mark Tasker for his help regarding albatrosses; Ray Tipper for his image of an immature Thick-billed Siskin; Vicente Valdés for the information and images shared; and Jhonson Vizcarra for his varied contributions and comments. Thank you very much to each and every one of you. I have surely forgotten someone, and I apologise for that. This project has taken many years and the memory or the accompanying notes are not always as they should be.

I wish to thank my friend and partner in Birding Chile, Juan Ignacio Zuñiga, for his company in the field, his ability to track hard-to-find birds, his support in publishing this English version of the book, and his permanent push to go a little further (both mentally and physically). Once again, thank you very much.

I also want to give my thanks to Daniel Martínez P. who is, in my opinion, the most talented bird illustrator in Chile, and with whom I have shared this project since its inception. We have endured long days in the field, byzantine desktop discussions and almost all our differences. Thank you for having invited me to participate in this project.

Finally, I greatly appreciate the support of Nigel Marven, who made it possible for this book to be published in English. Without his intervention, it is very likely that we would not have been able to make this guide known outside the scope of Chilean readers. Thank you so much for everything!

Gonzalo González Cifuentes

Dusky-tailed Canastero *Pseudasthenes humicola*, Lago Peñuelas National Reserve, Valparaíso Region, Chile (Jose F. Cañas).

FOREWORD

It's February 2018. A turquoise reservoir is bounded by steep mountain slopes, and the vault of the sky is as blue as a Dunnock's egg – although this country is Dunnock-free, so as blue as a Wren-like Rushbird's egg would be more apt. (You can see how blue this is on Plate 89 of this book.)

I'm in Chile, at an altitude of 2,500 metres, in the heart of the Andes. Gonzalo E. González Cifuentes – the author of this fabulous field guide – and I are near the El Yeso Dam. We're squelching through a bog, scanning the edges of puddles and cushions of moss for a very special bird. A hummingbird zooms by, olive-brown above, with a glittering green throat and a broad black patch on its belly. Gonzalo instantly identifies it as a male White-sided Hillstar.

The hummer seems to want to point us in the right direction and flies right over the cryptic bird we're searching for. Hunkered down next to a mossy hummock, is one of the world's most peculiar waders – the Diademed Sandpiper-Plover. I drink in the detail of every feather, the bird's belly has fine black barring, its neck is a rich rufous, it has a white head-band and a drooping bill. The supporting cast of Andean birds that day included Least Seedsnipes, sporting a line of feathers down their throats like black neck-ties, colourful sierra-finches and drabber ground-tyrants. Gonzalo shared his top tips for identifying them all. 'You should write a field guide,' I said.

He already had, but at the time there was only a Spanish edition. As I said goodbye, Gonzalo gave me a copy of the book as a gift. I was lecturing on a cruise around South America at the time, and that night in my cabin, I pored over the mouth-watering plates painted by Daniel E. Martínez Piña. I'd seen many of Chile's 468 species of birds, but still hadn't caught up with either of the huet-huets, or two of the country's six species of cormorants – the exquisite Red-legged and Guanay.

A few days later, I connected with the wanted duo that fish for a living. Guanay and Red-legged Cormorants flew in formation next to a fishing boat outside the port of Valparaíso. I was on board with Gonzalo, heading out to the Humboldt current, an upwelling of cold water laden with nutrients. This fertile zone is rich with marine life, especially seabirds.

Thirty miles out, the book really came into its own. As Gonzalo pointed out albatrosses, petrels and shearwaters, I checked out their features in the field guide. The seabird plates are some of the best I've ever seen. With the comprehensive paintings of bill patterns and plumage changes as the birds age, you can identify Chile's albatrosses with confidence. On that pelagic trip, we saw Northern Royal, Buller's and Salvin's. I also had my best-ever sightings of White-chinned and Westland Petrels – it was as if they'd flown off the page in the field guide. The birds were so close, I could compare the pale bill-tips of the White-chinned with the black-tipped bills of the Westland.

I'm thrilled that this superb field guide is now available in English. I'm sure it will get the readership it deserves and will be an invaluable companion on any trip to Chile. Flicking through the pages of the book at home in Somerset brings back so many memories. My copy is annotated with the Chilean birds I've seen already, but I still have to catch up with those huet-huets.

Nigel Marven

Crag Chilia *Ochetorhynchus melanurus*, La Serena countryside, Coquimbo Region, Chile (Gonzalo E. González Cifuentes).

INTRODUCTION

The book includes the birds that occur in continental and insular Chile, the country's Antarctic territory, as well as its territorial seas and Exclusive Economic Zone, using the criteria of the Chilean Ministry of Foreign Affairs, Boundaries and Borders. This is plotted on page 15, and includes the administrative divisions of the country as well as the localities mentioned in the texts, while the map on page 16 displays the main biogeographical zones of Chile.

SPECIES INCLUDED

This book describes a total of 468 bird species, including all those recorded at least five times in Chilean territory, based on criteria adopted by the authors. Birds recorded on fewer occasions are described as Extreme Vagrants on pages 188–191. Some other records are difficult to explain (such as Spotted Rail *Pardirallus maculatus*, Oilbird *Steatornis caripensis*, Common Nighthawk *Chordeiles minor*, Baltimore Oriole *Icterus galbula* or Swallow Tanager *Tersina viridis*). Others most likely emanate as escapees from captivity (e.g. Ruddy Shelduck *Tadorna ferruginea* or Red-crested Cardinal *Paroaria coronata*), and some have been observed only once, albeit by comparatively many birdwatchers (such as Scarlet-headed Blackbird *Amblyramphus holosericeus* or Whistling Heron *Syrigma sibilatrix*). Still others probably occur more frequently, but in areas of the country with few experienced observers, as is the case for most of Chile's offshore waters (examples include Flesh-footed Shearwater *Ardenna carneipes* and Great-winged Petrel *Pterodroma macroptera*). We also consider that some claimed occurrences reflect identification errors (as could have occurred in the cases of Laughing Falcon *Herpetotheres cachinnans*, Mountain Velvetbreast *Lafresnaya lafresnayi*, Slender-billed Miner *Geositta tenuirostris* or White-headed Steamerduck *Tachyeres leucocephalus*, all of which have been sometimes mentioned for Chile). Such records will not be eligible for full treatment in this guide until the minimum five records is reached.

Diademed Sandpiper-Plover *Phegornis mitchellii*, El Yeso Valley, Metropolitan Region, Chile (Gonzalo E. González Cifuentes).

Burrowing Parrot *Cyanoliseus patagonus*, close to La Serena, Coquimbo Region, Chile (Gonzalo E. González Cifuentes).

TAXONOMY

Field guides are not taxonomic authorities, but tools to help birdwatchers identify species in the field. Consequently, they should not attempt to 'resolve' taxonomic discussions or seek to be 'definitive' in this respect. Such work belongs to the scientific community. Nevertheless, at the time of publication, the available taxonomic information has been used (assuming some of the existing positions), in addition to our own criteria and field experience (whether rightly or wrongly). Genetic studies, and their refinement in terms of technology and types of molecular markers, have generated a significant amount of data aimed at resolving species limits and reordering the taxonomic tree. These have translated into changes in the organisation of families, sometimes yielding new arrangements within them, and the substitution of some genera for others (new, or unused for many years). This process is continuously ongoing, and an absolutely definitive version is unlikely to be reached in the near future. New information was included up until December 2019 and the main sources consulted are detailed in the Common (English) and scientific names section on page 12.

STRUCTURE OF THE BOOK

The book comprises plates, and short texts which summarise information about each species. The organisation of the plates and the birds differs from most field guides (which are based on taxonomy). Basing our work on the assumption that identifying a specific bird will be made easier if the illustrations bunch together the different species by the environments they inhabit, the plates have been broadly grouped into three categories: seabirds, waterfowl and landbirds. Of necessity, this breaks the taxonomic sequence that is common in books such as this.

Seabirds are mainly pelagic species, either offshore or linked to the continental shelf. Also included in this group are coastal birds, which depend on the ocean for their feeding grounds, although some may move inland, including several gulls and Snowy-crowned Tern *Sterna trudeaui*. It bears mention that some species present problems of classification in this respect, e.g. Seaside Cinclodes *Cinclodes nigrofumosus*, which despite being restricted to coastal environments is considered a landbird, thereby grouping it with other cinclodes and facilitating their comparison in the field.

Waterbirds inhabit and feed in non-marine wetlands of all types, and some could be considered seabirds, e.g. several sandpipers, plovers and Magellanic Steamerduck *Tachyeres pteneres*, while others that are generally found on freshwater bodies can also occur on the sea, for example grebes.

Seaside Cinclodes *Cinclodes nigrofumosus*, Zapallar Coast, Valpariso Region, Chile (Gonzalo E. González Cifuentes).

Landbirds include those that do not depend directly on the sea or wetland habitats, although there are still a few species that are more difficult to classify, such as Ringed Kingfisher *Megaceryle torquata* or those passerines that are closely associated with reeds in freshwater wetlands, such as Wren-like Rushbird *Phleocryptes melanops* and Many-coloured Rush-Tyrant *Tachuris rubrigastra*, or others such as the previously mentioned Seaside Cinclodes.

Finally, with respect to each plate, two main criteria have been adopted: to gather together species that can be confused with one another (for an example, see Plate 4), or to compare species that share a geographical area, and are therefore very likely to be seen together (as is the case with those species on Plate 10).

PLATES Within each plate the species are shown to relative scale. Most plates include up to a maximum of eight species (one shows nine), and illustrate both male and female of each species, as well as a juvenile or immature. When there are no differences between the male and female, a single adult is shown. In some cases chicks are illustrated, and for others more than one subspecies is depicted. In some plates, details of certain feathers, the wings or tail are presented, when it is considered that these are particularly important for field identification (as in the case of snipes' tails, Plate 42). Numerous small images depict different species within their typical environments, or show characteristic behaviour, often in comparison to other species, and giving the impression of distant field views. These smaller images are not always identified individually, giving the reader the opportunity to apply the species' identification criteria.

TEXTS These are highly standardised. For each species, the text includes common (English) and scientific names, overall length, wingspan (where appropriate), elevational range (if relevant), status, ID (focused on species that it can be confused with), habitat, and where to see it, alongside a distribution map.

We decided to reduce physical descriptions of the birds, to let the images 'do the talking', thereby providing more space for the texts to discuss other aspects of each species useful to the identification process (such as vocalisations), which are impossible to convey visually.

Common (English) and scientific names These follow mainly the South American Classification Committee (SACC) of the American Ornithological Society (AOS), and the *HBW and BirdLife International Illustrated Checklist of the Birds of the World* (in two volumes), or in the case of some albatrosses, the Agreement on the Conservation of Albatrosses and Petrels (ACAP). Occasionally, some other sources have been employed for scientific names.

Length (L) The overall length based on various literature sources, expressed in cm.

Wingspan (W) Presented only for those birds usually seen in flight, such as raptors and pelagic birds, also expressed in cm.

Elevational range Also referred to as altitudinal range, expressed in metres above sea level. The elevational ranges of some species vary geographically. These are based on the authors' own records and the available literature.

For the purposes of this guide, Chile has been subdivided into four areas: north (**N**), central (**C**), south (**S**) and austral (**A**).

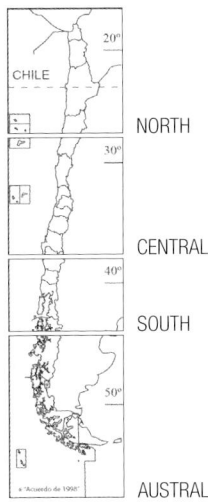

NORTH

CENTRAL

SOUTH

AUSTRAL

Sexual dimorphism Use of male (♂) and female (♀) symbols indicates whether the sexes differ or not.

Status Three main categories and the same number of subcategories are employed. Categories are assigned using a capital letter whereas the subcategories are designated by a lower case letter.

Main categories:

R: Resident. Species that breed in Chile, including permanent residents, e.g. Southern Lapwing *Vanellus chilensis*; migrants which to some extent vacate the country during the non-breeding season, for example Spectacled Tyrant *Hymenops perspicillatus*; and introduced species, e.g. California Quail *Callipepla californica*.

M: Migrant visitors. Species that visit Chilean territory annually, but do not breed there, making regular and well-documented migrations. It includes some species that can be found during any month of the year, either because some individuals do not leave Chile for their breeding grounds (for example, Whimbrel *Numenius phaeopus* or Lesser Yellowlegs *Tringa flavipes*) or because their non-breeding populations wander extensively prior to adulthood (e.g., Salvin's Albatross *Thalassarche salvini*).

V: Vagrants. Species recorded at least five times in Chilean territory, but not established as regular visitors or residents in the country (extreme vagrants with fewer records are outside the scope of this book). Presence in Chile is presumably casual, although future occurrences can to some extent be expected (e.g. Hooded Grebe *Podiceps gallardoi*).

Subcategories:

e: Endemic. A resident species whose breeding range is restricted to Chilean territory (for example Chilean Tinamou *Nothoprocta perdicaria*). Species with just a few records beyond the national borders are here maintained in this category to highlight their interest to birdwatchers.

i: Introduced. Species introduced to Chile by humans and that have formed stable resident populations, which would prove virtually impossible to eradicate (e.g., California Quail *Callipepla californica* or House Sparrow *Passer domesticus*).

?: Used if there is reasonable doubt concerning a species' current status, which is always explained (for example, Patagonian Tinamou *Tinamotis ingoufi*). Typically, both categories that might be appropriate for a given species are mentioned (e.g., Elegant Crested Tinamou *Eudromia elegans* **R/V?**).

ID The species with which confusion most frequently occurs in the field are indicated, along with the main plumage or other characteristics useful in the identification process.

Habitat The environments in which a species is most frequently observed. In some cases, these can be very restricted and precise (as in Seaside Cinclodes) but in others they are very broad (as in Dark-faced Ground-Tyrant *Muscisaxicola maclovianus*).

Voice The most common calls or songs given by a species are described, sometimes accompanied by an onomatopoeic transcription, based on the authors' interpretation and experience.

Where to see This section provides a very general indication of sites where the species is usually found (i.e. there is a good chance, but no guarantee, of seeing the bird in question). It is not intended to be exhaustive, but focuses on known and easily accessible places in Chile for birdwatching, such as national parks, estuaries, wetlands and other frequently visited sites.

Note For some species, additional information is given on taxonomy or status, or to clarify a particular point.

[Alt.] Some alternative English names are given.

DISTRIBUTION MAPS According to the range of each species, three reference maps are used: continental Chile, insular Chile, and South America. Different colours indicate seasonal and, to some extent, frequency status, as follows:

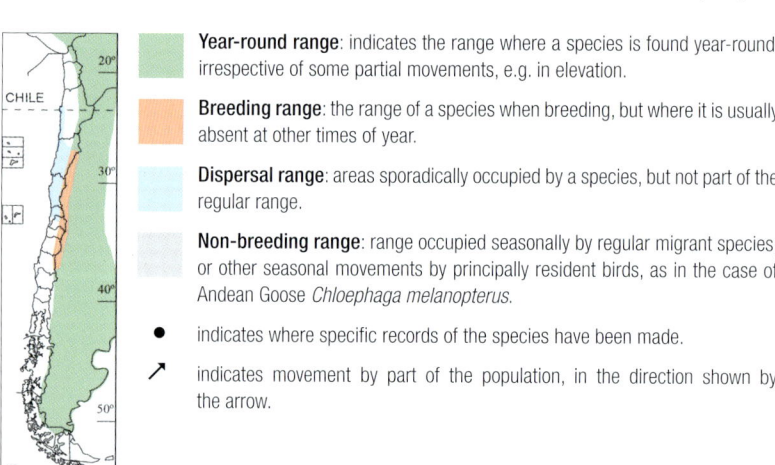

Year-round range: indicates the range where a species is found year-round, irrespective of some partial movements, e.g. in elevation.

Breeding range: the range of a species when breeding, but where it is usually absent at other times of year.

Dispersal range: areas sporadically occupied by a species, but not part of the regular range.

Non-breeding range: range occupied seasonally by regular migrant species, or other seasonal movements by principally resident birds, as in the case of Andean Goose *Chloephaga melanopterus*.

● indicates where specific records of the species have been made.

↗ indicates movement by part of the population, in the direction shown by the arrow.

ADMINISTRATIVE REGIONS OF CHILE

The approximate geographical location is indicated for locations, islands and other important areas mentioned in the species accounts, under the **Where to see** section.

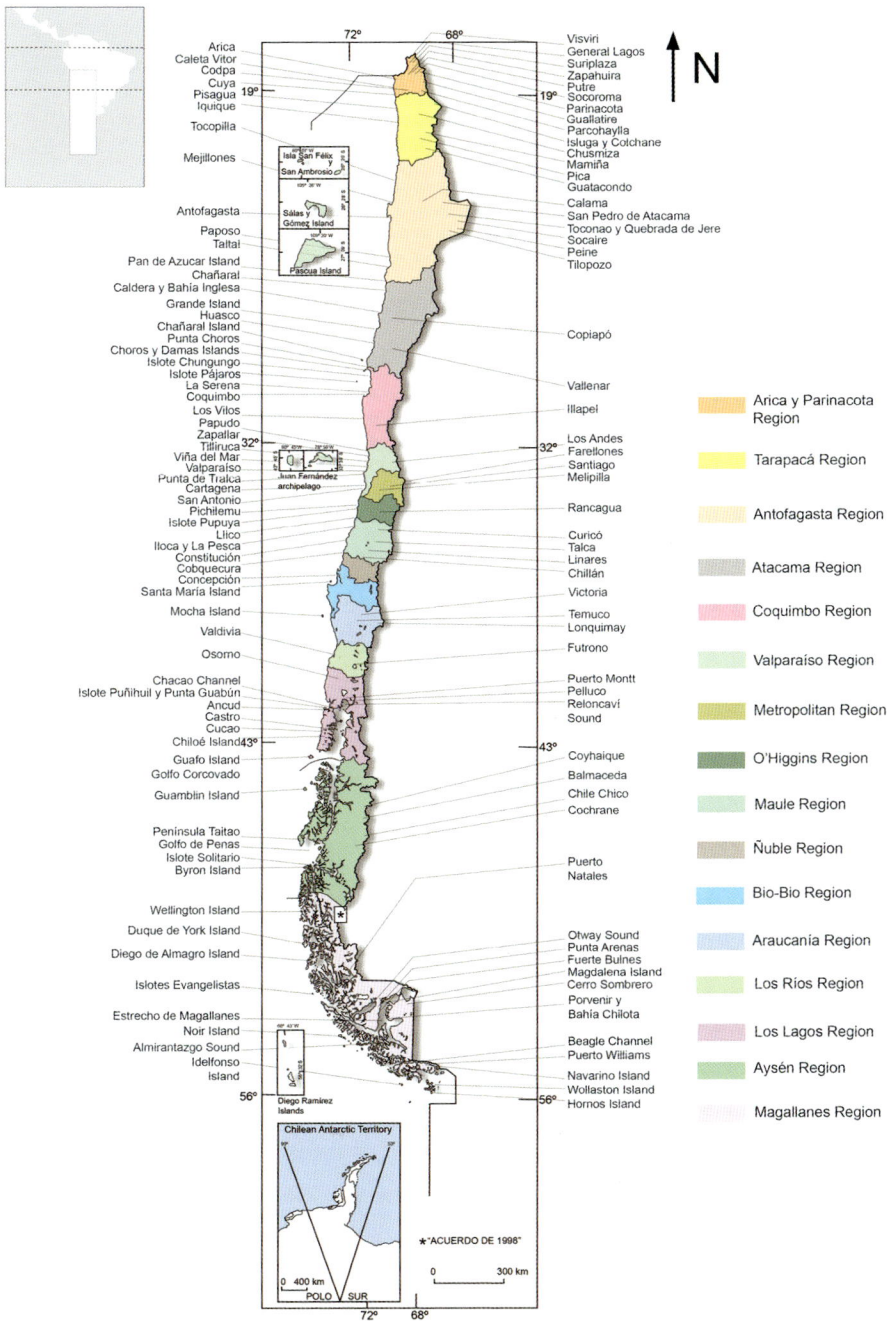

BIOGEOGRAPHIC ZONES OF CHILE

The approximate geographical location is indicated for sites of ornithological importance mentioned in the species accounts, under the **Where to see** section. Protected sites of interest are also indicated, and include the following names: National Park, National Reserve and Natural Monument.

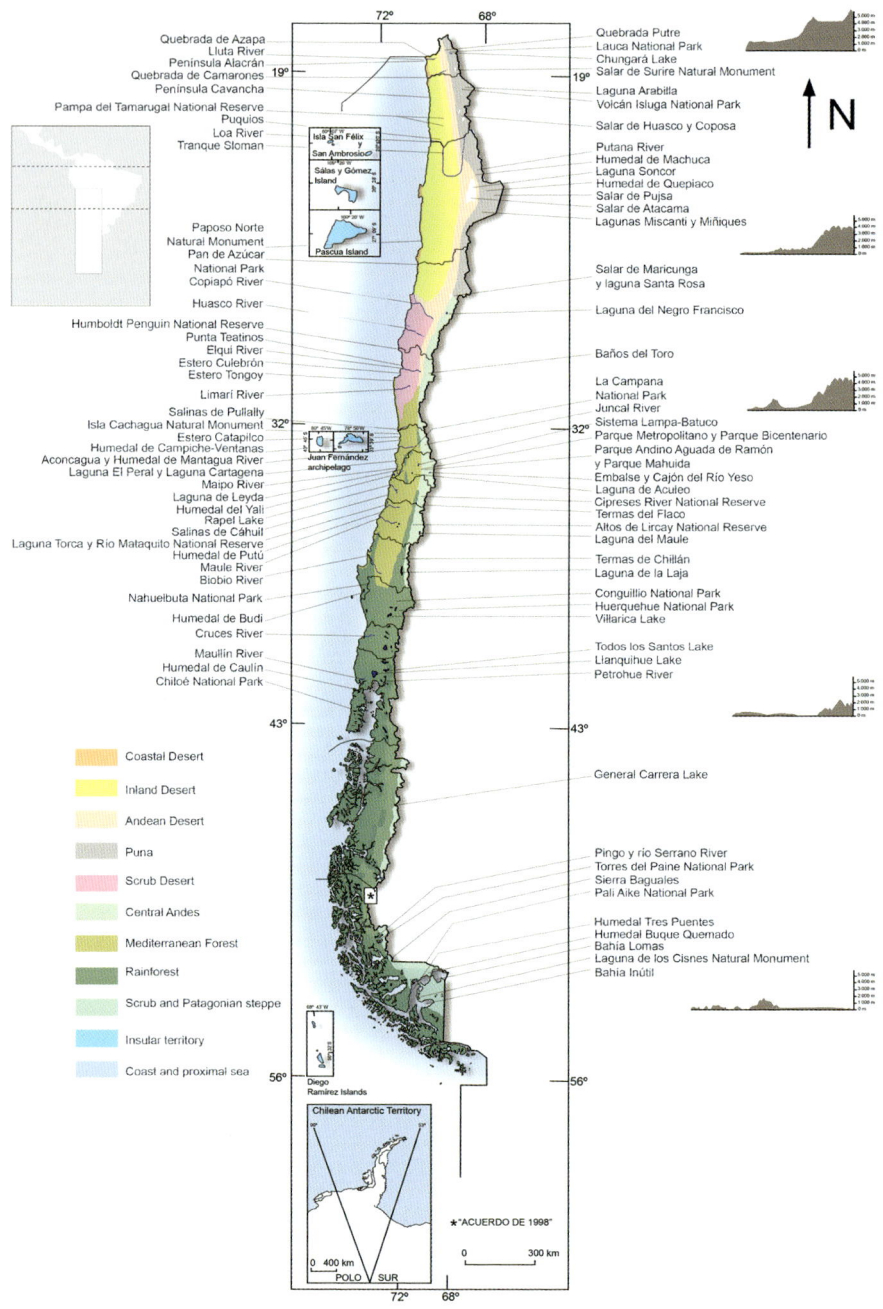

BIRD TOPOGRAPHY

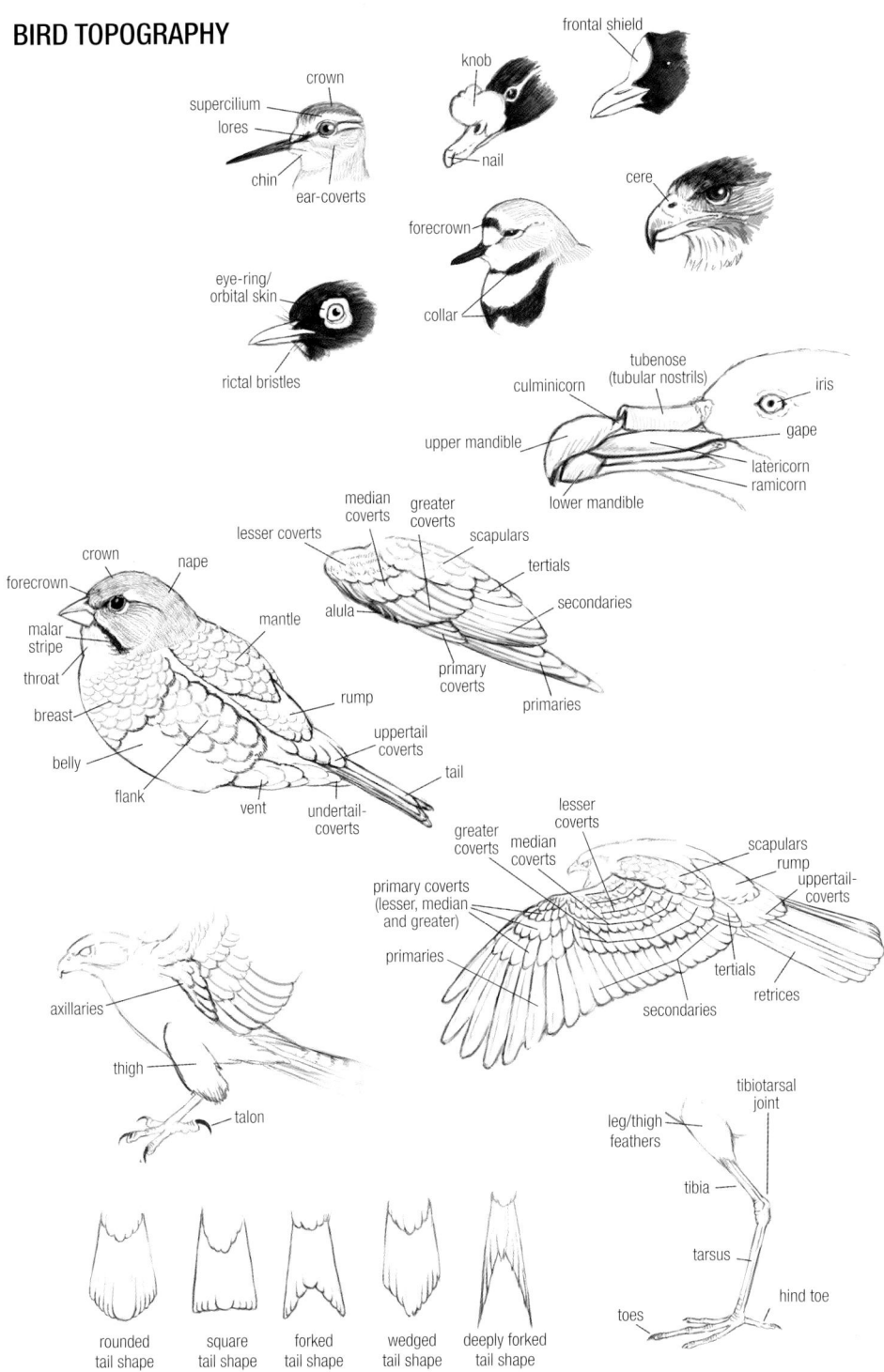

PLATE 1: PENGUINS I

Emperor Penguin
Aptenodytes forsteri   R
L 110–125cm. Sexes identical. **ID** Unmistakable. Largest penguin. An open yellowish patch behind ear-coverts becomes white towards the neck. **Habitat** Antarctica when breeding. Pelagic in non-breeding season. **Voice** Colonies are noisy and, of the several vocalisations, most characteristic of adult is a long buzzing *Ha-ha-ha-ha -...*, while another is sonorous and shorter, recalling a trumpet. Chicks emit a whistle-like call, plus a soft but acute melodious trill. **Where to see** Only Antarctica (e.g., Snow Hill Island, on the Antarctic Peninsula), where nests in colonies. No records on mainland.

King Penguin
Aptenodytes patagonicus   R
L 95–100cm. Sexes identical. **ID** Unmistakable. Has an orange patch behind the ear-coverts, which is very well demarcated, and can appear enclosed. **Habitat** Breeds on subantarctic islands, on sandy beaches, stony or with low vegetation. Pelagic in non-breeding season. **Voice** Colonies very noisy, and audible at long distance. Two characteristic vocalisations: a short, acute and melodious trill like a whistle, given softly by the chicks, and a long and sonorous call like trumpeting, which is typical of adults. **Where to see** Parque Pingüino Rey (Bahía Inútil, Magallanes Region), where a colony of c.100 birds is now established. Increasing records from coasts of Patagonia.

Chinstrap Penguin
Pygoscelis antarcticus   R
L 76cm. Sexes identical. **ID** Unmistakable. Note the thin black line surrounding the white throat. **Habitat** Antarctica. Also islands and subantarctic islets, to Cape Horn. Pelagic in non-breeding season. Occasional in channels in southernmost Chile. **Voice** Colonies noisy. Adults emit a kind of cackle, accelerating and becoming more aggressive towards the end. Chicks give a high-pitched whistle, sounding almost pitiful, and repeated several times. **Where to see** Antarctica. Can be seen on Deception Island, or at Las Estrellas, on King George Island (Magallanes Region).

Gentoo Penguin
Pygoscelis papua   R
L 76–80cm. Sexes identical. **ID** Unmistakable. Look for the white band over the crown and reaching a point above the eye, which is unique to this species. **Habitat** Subantarctic islands and Antarctica when breeding. Pelagic in non-breeding season. Also channels and islands in the far southern mainland. Rocky beaches. **Voice** Colonies are not very noisy. Adult voice recalls a braying, similar to Magellanic or Humboldt Penguins, but also gives trumpeting or cackling calls, always sounding high-pitched and shrill. Chicks emit high-pitched (but not shrill) whistles. **Where to see** Beagle Channel and Antarctica, e.g., at Las Estrellas, on King George Island (Magallanes Region).

Adelie Penguin
Pygoscelis adeliae   R
L 71cm. Sexes identical. **ID** Unmistakable. All-black head with a neat white eye-ring immediately identifies the species. **Habitat** Antarctica when breeding. Pelagic in non-breeding season. **Voice** A noisy penguin, especially at colonies. Most frequently heard call is a kind of trumpeting, hoarse and not very melodious. **Where to see** Antarctica, the only place where it nests, e.g. on King George Island (Magallanes Region).

PLATE 2: PENGUINS II

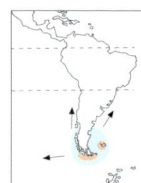

Southern Rockhopper Penguin
Eudyptes chrysocome   **R**
L 61cm. Sexes identical. **ID** Similar to Macaroni Penguin; look for double yellow crest (raised at will) and narrow yellowish eyebrow, starting almost at the base of the short dark red bill. **Habitat** Pelagic in the non-breeding season. Nests on steep-sided islands. **Voice** Noisy at colonies. Adult gives varied vocalisations, trumpeting, strident moans, and rough unmelodic groans. One call is likened to a mechanical tapping (*Kak-kak-kakkak-kak-kak-kroo...*) or scratching a metallic object. Chicks emit short, high-pitched calls, sounding pitiful, and given repeatedly when begging for food. **Where to see** Austral islands of very difficult access, e.g. Islote Solitario, in the Penas Gulf (Aysén Region), and Noir Island, Diego Ramírez Islands, or Ildefonso Islands, all in Magallanes Region.

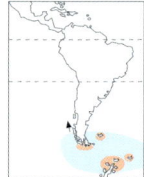

Macaroni Penguin
Eudyptes chrysolophus   **R**
L 70–76cm. Sexes identical. **ID** Similar to Southern Rockhopper Penguin. Has two orange/yellow plumes that meet above the bill, which is large and an intense garnet-red with pinkish gape. **Habitat** Less pelagic during non-breeding season than many penguins, preferring coasts and seas adjacent to islands where it breeds. Nests on steep-sided islands. **Voice** Very noisy at colonies, vocalising frequently, with a long voice similar to a cracking sound or pounding on plastic (*Kakakakakakakakaaaa...*), and a call similar to a cough (*Kho-kho-kho...*). Chicks utter very sharp whistles (*Kew…kew…kew…kew...*). **Where to see** Very difficult-to-access islands in the austral zone, like Duke of York Island, Noir Island, Diego Ramírez Islands or Ildefonso Islands, all in Magallanes Region.

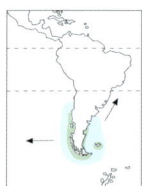

Magellanic Penguin
Spheniscus magellanicus   **R**
L 70cm. Sexes identical. **ID** Similar to Humboldt Penguin. At all ages, note the different insertion of the upper mandible in the face, compared to Humboldt Penguin. Immature shows no pink at bill base. Adult has two black breast-bands, and a broad white superciliary band encircling rear ear-coverts. **Habitat** Pelagic, but strongly associated with coasts and adjacent rocky islets when breeding. In the far south, on remote beaches, usually stony, and sometimes on small wooded islands. Nests in holes, requiring a sandy substrate for its colonies. **Voice** Very noisy around its colonies. Main vocalisation recalls a long moan ending in a braying, which gives rise to its Chilean name, 'Donkey duck' ('Pato burro'), which is also used for Humboldt Penguin. **Where to see** Islote Puñihuil (Los Lagos Region), Magdalena Island (Magallanes Region).

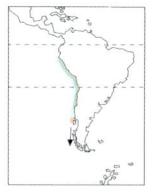

Humboldt Penguin
Spheniscus humboldti   **R**
L 70cm. Sexes identical. **ID** Similar to Magellanic Penguin. At all ages, note the different insertion of the upper mandible in the face, compared to Magellanic Penguin. In immature the base of the bill always shows some pink. Adult has a single black breast-band and thin white stripe emanating between the eye and the base of the bill, and curving behind the ear-coverts. **Habitat** Breeds on rocky coasts. Usually in coastal waters at other times. **Voice** Low-pitched vocalisations, clearly recalling a braying sound and similar to Magellanic Penguin. **Where to see** Pan de Azúcar National Park (Atacama Region), Humboldt Penguin National Reserve (Coquimbo Region), Pájaro Niño islet, Algarrobo, and Cachagua Island Natural Monument (Valparaíso Region).

Little Penguin
Eudyptula minor   **V**
L 40–45cm. Sexes identical. **ID** Adult unmistakable in good views, but at longer range could recall juvenile Magellanic and Humboldt Penguins. A very small penguin, with bluish-grey upperparts and pale yellow eyes. **Habitat** Pelagic in warm waters when not breeding. **Voice** Very vocal at colonies. Hoarse-sounding grunts, reminiscent of calls of Magellanic Penguin. Occasionally vocalises when swimming, at least around colonies, mainly at twilight and at night. **Where to see** Vagrants could show up anywhere in coastal zones.

PLATE 3: LARGE ALBATROSSES I

Identification in the field often very difficult due to subtle differences between some species and the number of intermediate stages in their plumage development, which are indistinguishable in some cases and often poorly known. The limits between these stages are unclear and imprecise, as plumage maturation represents a continuum, in which individual variation and different moulting patterns further complicate matters for birdwatchers. In the plates, four stages are defined: **juvenile**, or the first true plumage after the natal down is lost and in which birds leave the nest, which is maintained until c.6–18 months old, depending on species, after which they moult to the next stage; **early immature**, which corresponds to individuals aged between 6–18 months and 24–30 months, again depending on species; **late immature** plumage is retained until 48–60 months; and finally, **adult**, or basic definitive plumage, which is usually acquired after 60 months.

Northern Royal Albatross
Diomedea sanfordi **M**
L 110–120cm. **W** 320cm. Sexes identical. **ID** Generally similar to Southern Royal Albatross, and some stages are very similar, but latter slightly larger. Has basically all-black wings above at all ages (although some may show sporadic white mottling, which lacks any pattern). Juvenile shows large dorsal black spots, sometimes even a few blackish crown feathers. From below, at all stages note black border between the carpal and wingtip, which is usually (but not always) present, but never shown by adult Southern Royal. Both royal albatrosses show a black tomial line (along the bill's cutting edges) but never have pinkish markings on neck, separating them from Wandering Albatross. **Habitat** Pelagic in subantarctic and temperate waters. **Voice** Silent at sea (except when squabbling for food). Vocal during courtship displays at breeding colonies. **Where to see** Offshore, approximately between Valparaíso (Valparaíso Region) and the far south (Magallanes Region), including Drake Passage, but could occur anywhere in Chilean Humboldt Current. Does not follow fishing vessels, but may investigate them and settle on the water if there is chum or fish discards to be had.

Wandering Albatross
Diomedea exulans **M**
L 107–130cm. **W** 340–350cm. Sexes identical. **ID** Similar to Antipodean Albatross, with adult of latter very similar to immature stages of present species, as well as to Southern Royal Albatross. Immature Southern Royal has scapulars more darkly stained than Wandering, and these delimit a white dorsal area, which is rounded or elliptical in Southern Royal, but triangular or wedge-shaped in present species, because the scapulars are white. Wings acquire white from a point in the central third, and spreading forwards, whereas in Southern Royal the wings acquire white from the leading edge backwards. Tail of Wandering shows black feathers, even in young adult, and some display pinkish neck markings, which are never present in royal albatrosses. **Habitat** Pelagic in subantarctic, temperate and even tropical waters as far north as 22°S. **Voice** Vocalises when resting on sea, not only during disputes for food. Complex, very varied and evocative vocalisations accompany the elaborate courtship displays, along with bill-clapping. **Where to see** Usually in waters beyond the Humboldt Current, especially from La Serena south to the Drake Passage. Follows all kinds of vessels, sometimes for long periods and in numbers, squabbling for any food in the water. [Alt. Snowy Albatross]

Southern Royal Albatross
Diomedea epomophora **M**
L 107–122cm. **W** 330–350cm. Sexes identical. **ID** Similar to Northern Royal and Wandering Albatrosses. Juvenile has black wings with white-fringed coverts forming fine longitudinal lines, but with wear the wings become virtually all black, except for the white leading edge (Northern Royal occasionally shows this). Until the early immature stage has a black border between the carpal and wingtip, which is afterwards lost. Towards adulthood the wings turn white from the leading edge rearwards, whereas in Wandering the wing acquires white from the centre first. Other features that separate both royal albatrosses from Wandering are described under Northern Royal. **Habitat** Pelagic in subantarctic and temperate waters. **Voice** Silent at sea (except in disputes over food). Vocal during courtship displays and at the nest, with calls that recall a snarling pig or a Neotropic Cormorant, and forceful bill-clapping displays. **Where to see** Throughout Chilean waters, but concentrated south of the Penas Gulf (47°S) to the Drake Passage. Follows fishing boats and feeds on fish discards and chum.

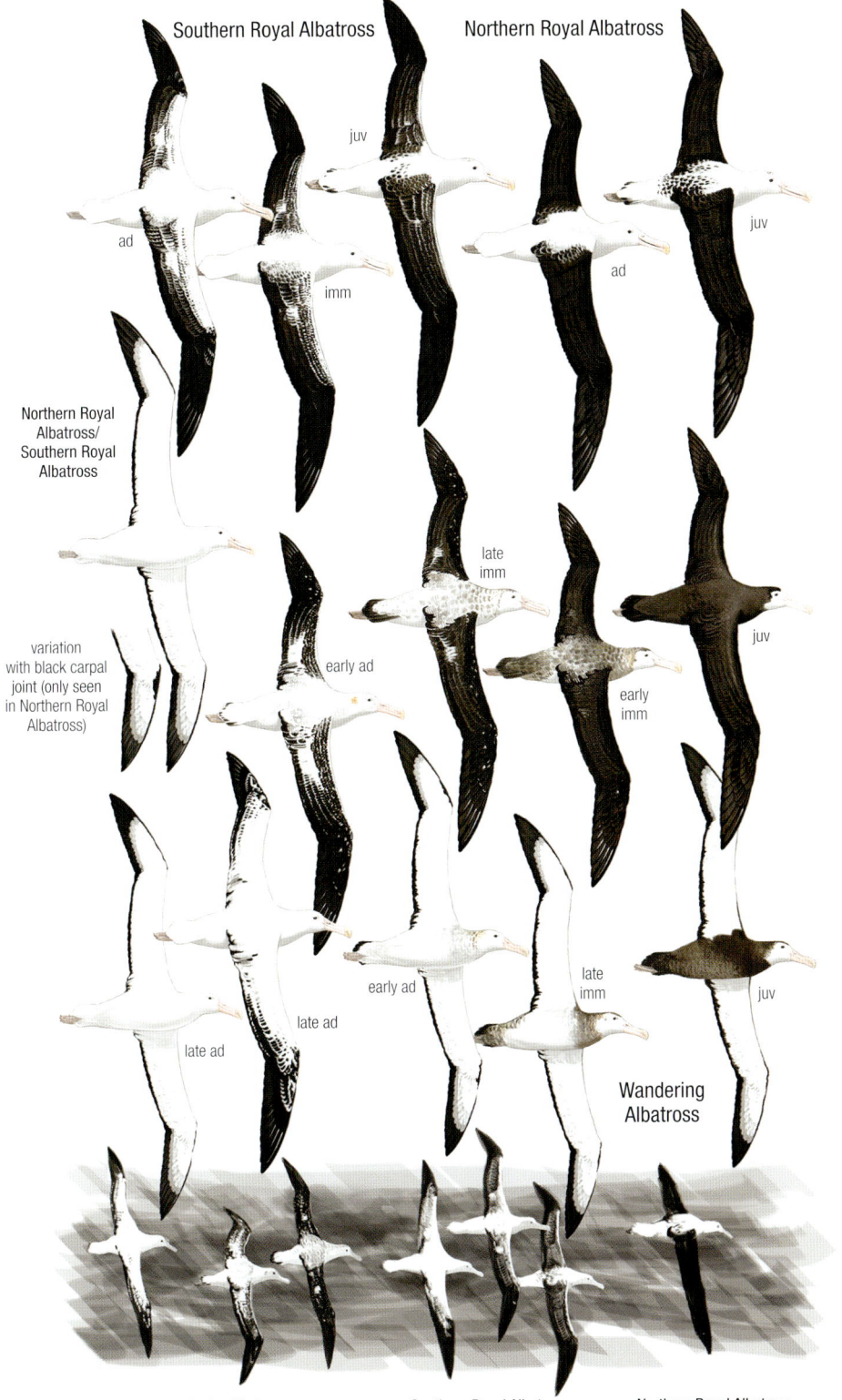

PLATE 4: LARGE ALBATROSSES II AND GIANT PETRELS

Antipodean Albatross
Diomedea antipodensis   **M**
L 110–117cm. **W** 280–335cm. Sexually dimorphic. **ID** Adult and juvenile similar to Wandering Albatross at sea, and practically indistinguishable in many plumage stages. Crown is brown and well defined, the wings dark above, and has black tail at all ages. **Habitat** Pelagic in subantarctic and temperate waters. **Voice** During elaborate courtship displays calls very varied, evocative and complex, and mixed with bill-clapping. **Where to see** From around Valdivia (39.8°S) south. Usually far offshore in region of the Taitao Peninsula (Aysén Region). Follows fishing vessels, sometimes for long distances, and is attracted to fish waste and chum.

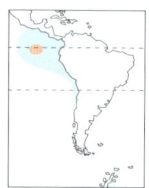

Waved Albatross
Phoebastria irrorata   **V**
L 89cm. **W** 235cm. Sexes identical. **ID** Unmistakable. Long-winged albatross, with long all-yellow bill and creamy-yellow nape. **Habitat** Only albatross that breeds in equatorial waters. Very restricted range, including open waters of continental shelf. **Voice** Silent at sea, but in courtship gives a series of whimpers and trumpet-like calls, as well as bill-clapping. **Where to see** Most likely offshore between Arica y Parinacota Region) and Antofagasta (Antofagasta Region). Does not follow large fishing vessels, but will cautiously approach small boats, and can feed on fish waste or chum.

Light-mantled Albatross
Phoebetria palpebrata   **M**
L 82–85cm. **W** 200–208cm. Sexes identical. **ID** Unmistakable in Chilean waters. Some young birds at long range appear very pale grey, even whitish, with darker wings. Note the long, sharply pointed tail. **Habitat** Pelagic, occupying Antarctic waters (south to 70°S) to tropical seas (close to 20°S, in Humboldt Current). **Voice** Silent at sea, except in aggression at food. Colonies are not especially noisy, but adult has disyllabic high-pitched calls, likened to braying, and also performs bill-clapping ceremonies. **Where to see** Generally only south of Valdivia (c.40°S). Rarely follows ships, and when does so typically loses interest quickly.

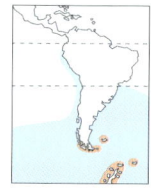

Southern Giant Petrel
Macronectes giganteus   **R**
L 88–92cm. **W** 205–222cm. Sexes identical. **ID** Similar to Northern Giant Petrel, and juvenile and immature stages are difficult to separate. Bill tip is greenish at all stages in Southern Giant. In poor light, the bill appears to be a single colour. White morph scarce (c.10% of population). **Habitat** Pelagic. In Antarctic (where nests), subantarctic and temperate waters, north to c.23°S, but sometimes reaches tropical waters following the Humboldt Current. Will approach the coast, especially in bad weather. Follows all types of boats, especially smaller fishing vessels, being attracted to any food in the water. **Voice** Generally silent, but gives aggressive grunts and guttural sounds at sea when competing for food, and at breeding colonies. Also bill-claps with considerable force. **Where to see** Humboldt Current, Chacao Channel (Los Lagos Region), Magellan Strait, Beagle Channel, Diego Ramírez Islands and Antarctic coasts (Magallanes Region).

Northern Giant Petrel
Macronectes halli   **M**
L 84–92cm **W** 205–222cm. Sexes identical. **ID** Like Southern Giant Petrel (which see). Bill tip reddish in all stages, in poor light appearing dark-tipped. **Habitat** Pelagic in subantarctic and temperate waters, occasionally reaching tropical waters (in the Humboldt Current) and the Antarctic Peninsula. **Voice** Generally silent, but gives aggressive grunts and guttural sounds when competing for food at sea, and at colonies. Forceful bill-clapping also heard. **Where to see** Along Humboldt Current, mostly from central Chile south, although present throughout, sometimes close to land. Chacao Channel (Los Lagos Region), channels in the far south (Los Lagos Region–Aysén Region), Magellan Strait, Beagle Channel (Magallanes Region).

PLATE 5: MOLLYMAWKS I

Black-browed Albatross
Thalassarche melanophris   **R**
L 80–90cm. **W** 229cm. Sexes identical. **ID** Adult unmistakable. Juvenile could be confused with juvenile Grey-headed Albatross and, under some conditions, Salvin's Albatross. Particularly confusing are juveniles with dirty grey heads, but note the typical bill coloration, grey with a dark tip in juvenile and immature, plus the elongated and well-pointed latericorn, which can facilitate identification from photos. **Habitat** Pelagic, mainly in subantarctic and temperate waters, but also Antarctic and tropical waters. Prefers the continental shelf, approaching coasts. Also in the southern channels. **Voice** Vocalises aggressively at sea when disputing food with other birds, emitting a long, nasal, single squawk (*Ghaaaaaaa...*). Very noisy at colonies, where gives a long, repeated pounding sound (*Gh-gh-gh-gh-gh-gh-gh-gh-gh...*) and a long, sharp, nasal moan (*Huuuueeeeee...*), as well as bill-clapping at variable intensity. **Where to see** The most frequently seen albatross throughout Chile, sometimes even from coasts, e.g. at Punta de Tralca (Valparaíso Region), Guabún (Los Lagos Region), Fuerte Bulnes road (Magellan Strait) and the Beagle Channel (Magallanes Region). Follows all types of the boats, sometimes in large numbers, especially fishing vessels, being attracted to fish waste and chum.

Grey-headed Albatross
Thalassarche chrysostoma   **R**
L 82–84cm. **W** 203–215cm. Sexes identical. **ID** Adult could be confused with Buller's and Pacific Albatrosses, but note the all-grey head and thick black border to the underwing. Juvenile and immature similar to those of Black-browed, Salvin's, Buller's and Pacific Albatrosses, but in juvenile note the concolorous blackish bill and shape of the latericorn, which distinguish it from Black-browed (a significant identification risk at sea). In immatures, grey on the head gradually increases and yellow appears on the bill. **Habitat** Pelagic, in subantarctic and temperate waters. Prefers open seas away from the continental shelf, but in bad weather may appear in channels in the far south. **Voice** Silent at sea, only vocalising during disputes over food with other birds, but at colonies is heard frequently, giving trumpeting sounds and moaning calls, as well as bill-clapping of variable intensity. **Where to see** Near colonies on Diego Ramírez Islands and Ildefonso Islands (Magallanes Region). More common in waters south of Valdivia (Los Ríos Region). Does not follow boats.

Buller's Albatross
Thalassarche bulleri bulleri   **M**
L 76–81cm. **W** 205–213cm. Sexes identical. **ID** Adult could be confused with Grey-headed and Pacific Albatrosses, the latter being almost indistinguishable. Juvenile and immature similar to Black-browed, Grey-headed and Salvin's Albatrosses, and indistinguishable from Pacific in same plumages. In adult note white underwing with broad black edges, and pale grey head with a dark area around eyes that does not reach the bill base. In juvenile bill appears long, rather thin, pinkish-grey with dark tip and well-defined plates, and same underwing pattern of adult. **Habitat** Pelagic, mainly in temperate waters, but can reach tropical seas. **Voice** Silent at sea, vocalises aggressively only when disputing food with other birds. At colonies, mainly gives a very low, descending cackle (*Kho-kho-kho-kho-khooooo...*), as well as a long, shrill monosyllable with a sad timbre, similar to other albatrosses, and several calls between trumpets and sharp grunts with a metallic tone, vaguely reminiscent of a Neotropic Cormorant. **Where to see** Humboldt Current, principally between La Serena (Coquimbo Region) and Valdivia (Los Ríos Region). Follows fishing vessels of all types, taking fish waste and chum. [Alt. Southern Buller's Albatross]

Pacific Albatross
Thalassarche bulleri platei   **M**
L 85cm. **W** 210–213cm. Sexes identical. **ID** Adult could be confused with same-age Grey-headed and Buller's Albatrosses, and latter almost indistinguishable. Juvenile and immature similar to Black-browed, Grey-Headed and Salvin's Albatrosses, while Buller's Albatross is indistinguishable prior to adulthood. In adult, note white underwing with broad black edges, and bill size, shape and coloration (compared to Grey-headed), which is slightly broader, deeper and shows more black than Black-browed, with a white forehead on otherwise dark grey head, and an 'angry' appearance. Juvenile and immature have a long, rather narrow, dark-tipped pinkish-grey bill with well-defined plates, and same underwing pattern as adult. **Habitat** Pelagic, mainly in temperate waters, but also reaches tropical waters. **Voice** Silent at sea. Calls frequently at colonies, emitting high-volume moans, followed by a trumpeting sound, sometimes high-pitched trumpets alone. **Where to see** Humboldt Current, principally between La Serena (Coquimbo Region) and Valdivia (Los Ríos Region). Follows fishing vessels of all types, taking fish waste and chum. **Note** The two subspecies of *T. bulleri* require further study, given clear morphological differences and well-isolated colonies (1,340km apart). They perhaps represent separate species and are treated separately here. [Alt. Northern Buller's Albatross]

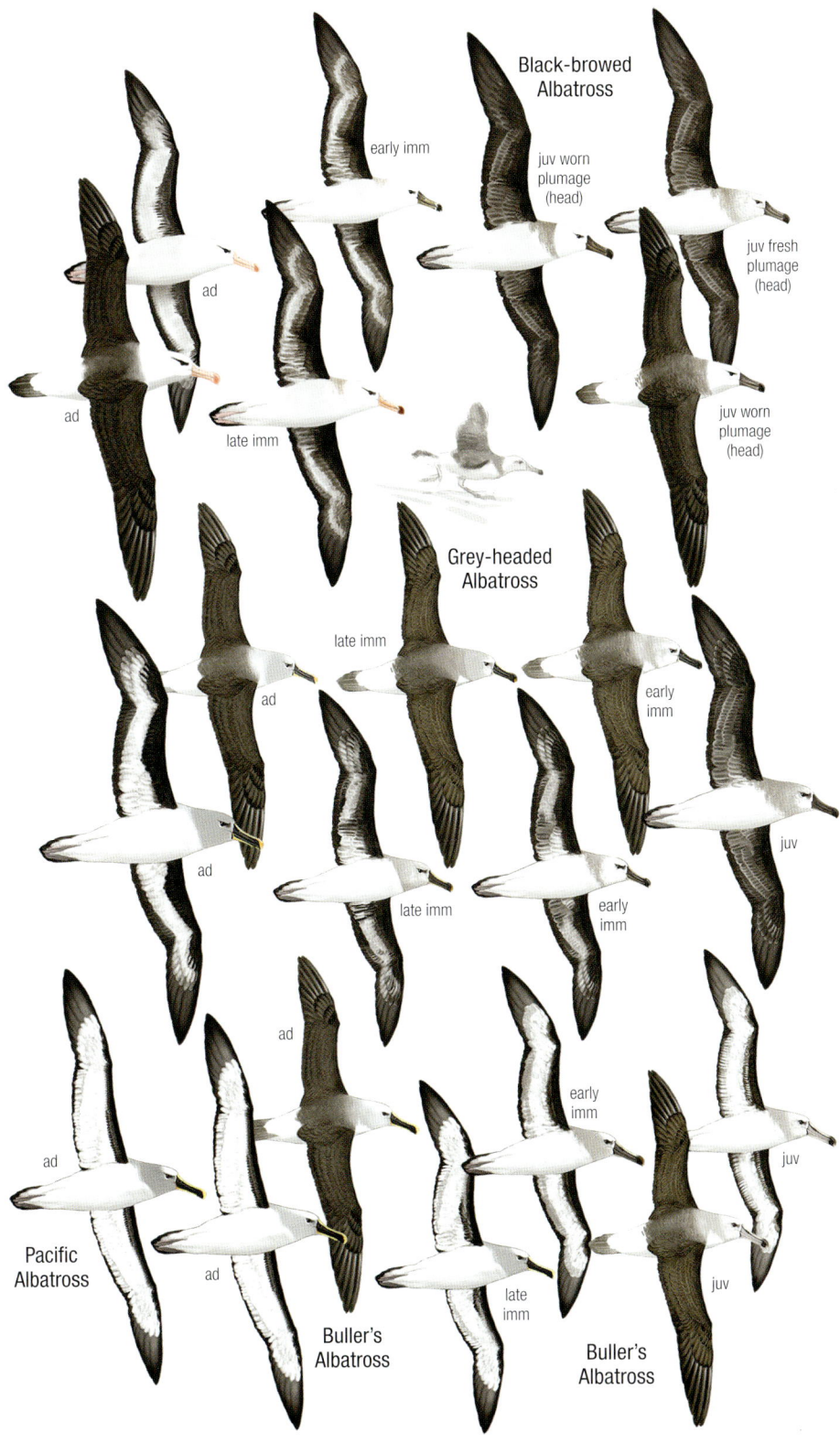

PLATE 6: MOLLYMAWKS II

The following three species were considered subspecies of Shy Albatross *T. cauta*, but the group has now been reclassified as four different species. They are difficult to separate; all have a blackish lores, giving them a unique appearance. The underwings are white, with a narrow black border and a rounded area at the junction of the leading edge with the body, which develops in the second year of life. Usually found a few km offshore in Chile, in the Humboldt Current.

Salvin's Albatross
Thalassarche salvini   **M**
L 90cm. **W** 250cm. Sexes identical. **ID** Adult unmistakable. Juvenile and immature similar to those of Black-browed, White-capped and Chatham Albatrosses. Note the white underwing with a thin black border (compared to Black-browed), grey bill with pale yellowish hue and dark tip (compared to Chatham) and the pale greyish head (compared to White-capped). **Habitat** Pelagic, mainly in temperate waters, but does reach tropical seas following Humboldt Current. Records between c.7°S and 42°S. **Voice** Silent at sea, except when disputing food with other birds, when notably aggressive. Noisy at colonies, mostly uttering trumpeting sounds and grunts, nearly always shrill and with a timbre similar to a toy trumpet. **Where to see** Humboldt Current, between northern limit and c.42°S (Los Lagos Region). Probably the second most numerous albatross in Chile, after Black-browed. Follows fishing vessels of all types, sometimes in large numbers, taking fish waste and chum.

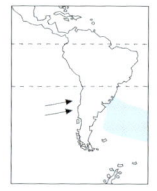

White-capped Albatross
Thalassarche steadi   **V**
L 94–100cm. **W** 244–260cm. Sexes identical. **ID** Adult unmistakable in Chile. Largest mollymawk. Juvenile and immature similar to those of Black-browed, Salvin's and Chatham Albatrosses. Underwing white with a thin black border, unlike Black-browed, and colour of head and bill separate it from Salvin's and Chatham. Culminicorn base grey. **Habitat** Pelagic, mainly in temperate waters, but may reach tropical seas following the Humboldt Current. **Voice** Silent at sea (except when disputing food). Vocalisations at colonies like others in this group. **Where to see** Humboldt Current, between northern limit and c.50°S. Very rare off central Chile (Valparaíso Region). Follows fishing boats, taking fish waste and chum. **Note** Satellite tracking (ACAP data) shows that *T. steadi* reaches Chile and the fourth species in this group (Shy Albatross *T. cauta*) disperses west from New Zealand.

Chatham Albatross
Thalassarche eremita   **M**
L 90cm. **W** 220cm. Sexes identical. **ID** Adult unmistakable. Juvenile and immature similar to those of Black-browed, White-capped and Salvin's Albatrosses, especially first-years. At distance, the white underwing with a thin black border separates it from Black-browed. First-year has bill like Salvin's, but head and neck darker grey, though not always obvious in the field. Immature acquires variably yellow-orange bill, which together with grey head and neck separates it from Salvin's and White-capped. **Habitat** Pelagic, mainly in temperate waters, but may reach tropical seas following the Humboldt Current. **Voice** Silent at sea (except at disputes over food). Noisy at (sole) colony, giving trumpeting sounds, as well as diverse high-pitched and shrill calls. **Where to see** Humboldt Current, mostly between northern limit and c.40°S. Follows fishing vessels of all types, sometimes in large numbers, taking fish waste and chum.

PLATE 7: ALBATROSS HEADS

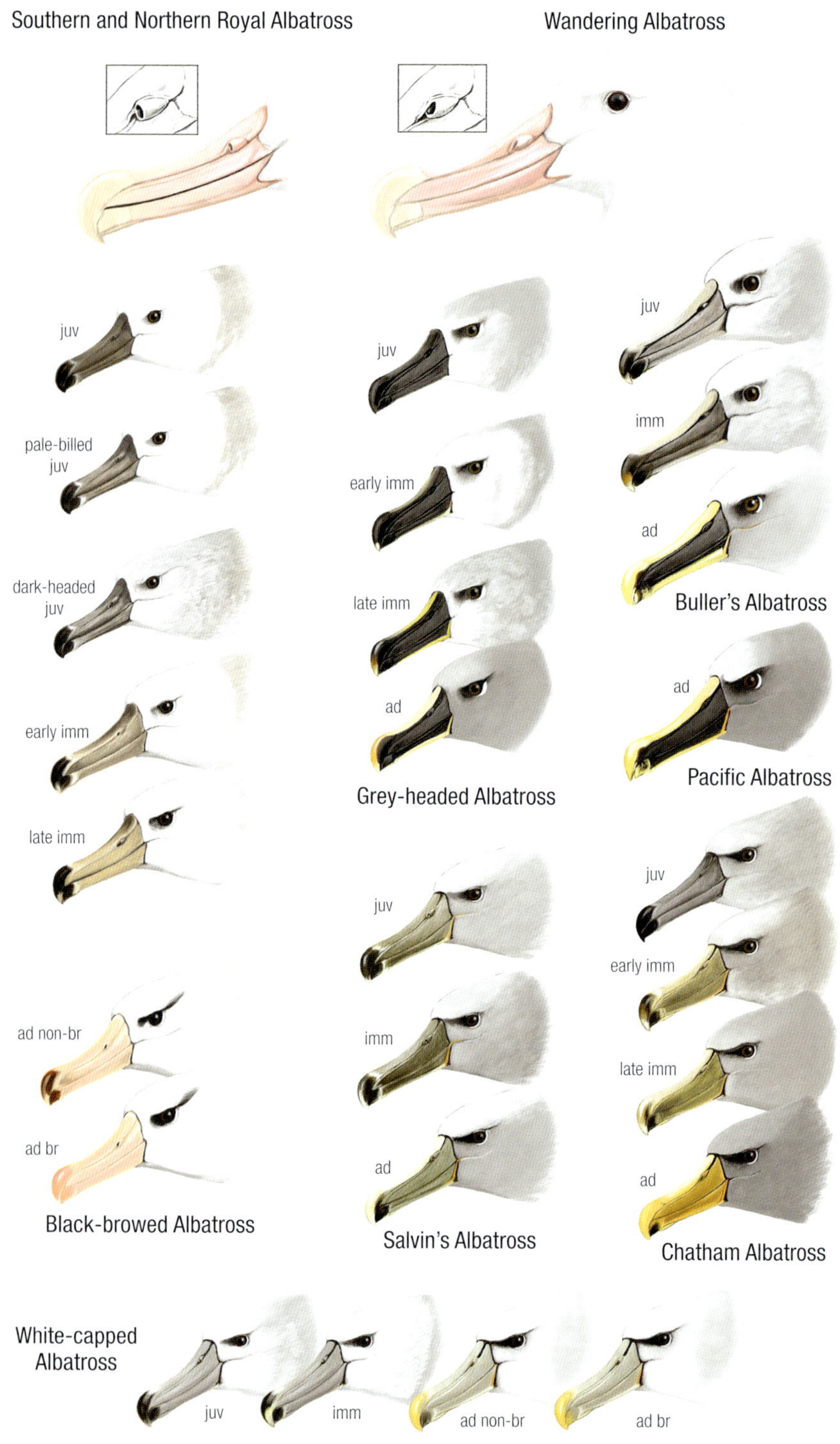

Southern and Northern Royal Albatross

Wandering Albatross

juv

pale-billed juv

dark-headed juv

early imm

late imm

ad non-br

ad br

Black-browed Albatross

juv

early imm

late imm

ad

Grey-headed Albatross

juv

imm

ad

Salvin's Albatross

juv

imm

ad

Buller's Albatross

ad

Pacific Albatross

juv

early imm

late imm

ad

Chatham Albatross

White-capped Albatross

juv

imm

ad non-br

ad br

PLATE 8: ALBATROSSES AT REST

PLATE 9: PETRELS AND SHEARWATERS I

Southern Fulmar
Fulmarus glacialoides R
L 45cm. **W** 120–130cm. Sexes identical. **ID** Unmistakable. Largely grey-and-white robust-bodied petrel, with dark-tipped pink bill, dark trailing edge to wing and obvious pale patch in otherwise dark primaries. **Habitat** Pelagic, mainly in Antarctic, subantarctic and temperate waters, but occasionally reaches tropical seas via the Humboldt Current. **Voice** Silent at sea, calling only when disputing food. Very noisy at colonies, especially early in the season, giving high-pitched, shrill and repetitive calls. **Where to see** Humboldt Current and Drake Passage, being attracted to fish debris and chum, and following all manner of vessels.

Cape Petrel
Daption capense R
L 40–41cm. **W** 91cm. Sexes identical. **ID** Unmistakeable. Unique black-and-white pattern. **Habitat** Pelagic, in Antarctic, subantarctic, temperate and tropical waters, following Humboldt Current. **Voice** Silent at sea, except when disputing food. Various calls (cackles, growls, purrs and shrieks) heard at colonies, high-pitched and at some volume. **Where to see** Throughout country, in Humboldt Current. Follows all types of fishing vessels, attracted to fish debris and chum.

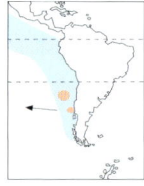

Pink-footed Shearwater
Ardenna creatopus R
L 45–48cm. **W** 99–109cm. Sexes identical. **ID** Unmistakable. Dark morph (very rare) can be confused with other dark shearwaters, but note pinkish bill with black tip. **Habitat** Pelagic but always close to coasts. **Voice** Sometimes calls in flight, but mostly in disputes over food, where can be noisy. On arriving at nest calls are nasal but very strong (*Peeeweeeee... Peeeweee…weeee…*), producing a cacophony of sound at large colonies. **Where to see** Humboldt Current in spring and summer; also Mocha Island (Araucanía Region), and Robinson Crusoe and Santa Clara in Juan Fernández archipelago (Valparaíso Region). Follows small fishing boats, attracted to fish waste and chum.

White-chinned Petrel
Procellaria aequinoctialis M
L 50–57cm. **W** 132–145cm. Sexes identical. **ID** Could be confused with Westland Petrel. Note all-yellowish bill. White chin not usually visible. **Habitat** Pelagic, mainly in subantarctic and temperate waters, but also tropical seas. Will approach coasts. **Voice** Silent at sea, except at disputes over food. Soft, trilled calls during non-aggressive social interactions, among flocks resting on sea. At colonies, most characteristic is a noise likened to an old sewing machine in a cobblers, which gave rise to its name in the Falkland Islands (Shoemaker). **Where to see** Humboldt Current. Follows all manner of vessels, including small fishing boats, attracted to fish waste and chum.

Westland Petrel
Procellaria westlandica M
L 51cm. **W** 136cm. Sexes identical. **ID** Compared to White-chinned Petrel, note yellowish bill has a black tip. **Habitat** Pelagic, in temperate and even tropical waters of the Pacific, concentrated in the Humboldt Current. **Voice** Silent at sea, except when disputing food, but very vocal at colonies, where most calls are generally nasal and shrill. **Where to see** Humboldt Current, following all types of fishing vessels, attracted by fish debris and chum.

Sooty Shearwater
Ardenna grisea R
L 43–46cm. **W** 97–106cm. Sexes identical. **ID** Similar to Christmas Shearwater (Plate 11). Very dark, with underwing pale grey to whitish (exhibiting contrast in flight); bill and legs black. Often in large flocks. **Habitat** Pelagic, in Humboldt Current, but even in bays or open channels in southern Chile. Regular in large flocks just offshore. **Voice** Silent at sea, except if competing for food when gives harsh calls. At colonies vocalises constantly, both in flight and from burrows. **Where to see** Humboldt Current. Easy in spring in Chacao Channel (Los Lagos Region) and Beagle Channel (Magallanes Region). Does not follow boats, but can be attracted to chum.

Buller's Shearwater
Ardenna bulleri M
L 43–46cm. **W** 96–104cm. Sexes identical. **ID** Unmistakable. Resembles a *Pterodroma* petrel in coloration, but flies differently. **Habitat** Pelagic; Pacific Ocean, mainly in Northern Hemisphere. **Voice** Silent at sea, but noisy at colonies. **Where to see** Humboldt Current, especially north of La Serena. Follows fishing vessels of all types, attracted by fish debris and chum.

Great Shearwater
Ardenna gravis M/V
L 45–49cm. **W** 108–116cm. Sexes identical. **ID** Unmistakable; dark cap very distinctive. **Habitat** Pelagic, mainly in Atlantic Ocean. Off Chile, apparently follows the Humboldt Current. **Voice** Vocalises at sea, when resting in flocks. Especially noisy at colonies, with sexually dimorphic calls. **Where to see** Most likely in Strait of Magellan, Beagle Channel or Cape Horn (Magallanes Region). Follows fishing vessels, attracted by fish debris and chum.

PLATE 10: PETRELS AND SHEARWATERS II

Stejneger's Petrel
Pterodroma longirostris R
L 29–32cm. **W** 70–74cm. Sexes identical. **ID** Similar to Juan Fernandez and Masatierra Petrels. Dirty white forehead reaches forecrown. Rest of crown and sides of face blackish, becoming grey on mantle and neck-sides. Black tail. Bill short and thin. **Habitat** Pelagic, in temperate and tropical waters of the Pacific. **Voice** Silent at sea; only vocalises during nocturnal visits to colonies. **Where to see** Breeds on Alejandro Selkirk Island, Juan Fernández archipelago (Valparaíso Region). Not strongly linked to Humboldt Current. Does not follow vessels or fishing boats, and is not especially attracted to fish waste or chum.

Juan Fernandez Petrel
Pterodroma externa R
L 42–45cm. **W** 101–112cm. Sexes identical. **ID** Similar to Stejneger's and Masatierra Petrels. Whitish tail base (not always visible), white forehead, dark crown that covers eye and is separated from mantle by pale collar. Black tail, in some with white tip. Bill short and robust. **Habitat** Pelagic, in temperate and tropical waters of eastern and central Pacific. Nests on oceanic islands. **Voice** Silent at sea, but utters various, not especially loud, calls at colonies. **Where to see** Breeds on Alejandro Selkirk Island, Juan Fernández archipelago (Valparaíso Region). Not strongly linked to Humboldt Current. Occasionally follows fishing boats, but rarely attracted to them, and will only briefly inspect smaller craft.

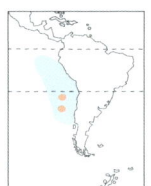

Masatierra Petrel
Pterodroma defilippiana R
L 26cm. **W** 70–76cm. Sexes identical. **ID** Similar to Cook's, Stejneger's and Juan Fernandez Petrels. Dark patch around eye with short white eyebrow. Head and upperparts uniform grey, and has a grey tail without dark tip. Bill long and robust. **Habitat** Pelagic, in temperate and tropical waters of eastern Pacific. Not strongly linked to Humboldt Current. **Voice** Silent at sea, but very noisy and diverse vocalisations heard at colonies. **Where to see** Around Juan Fernandez archipelago, especially Alejandro Selkirk Island, also San Félix and San Ambrosio, in Desventuradas Islands (Valparaíso Region). [Alt. De Filippi's Petrel]

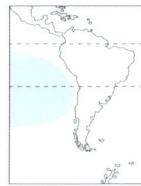

Cook's Petrel
Pterodroma cookii M
L 31–34cm. **W** 76–82cm. Sexes identical. **ID** Similar to Masatierra Petrel. Dirty white forehead, small black patch around eye and thin white eyebrow, which is not always visible. Also a fine black terminal spot on central tail feathers (also not always visible). Bill long and slim. **Habitat** Pelagic, in Pacific, from New Zealand to Aleutian Islands. **Voice** Vocalises at sea in disputes over food. **Where to see** No information.

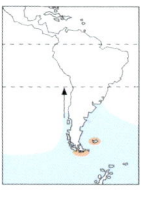

Blue Petrel
Halobaena caerulea R
L 28–30cm. **W** 48–58cm. Sexes identical. **ID** Unmistakable. Grey tail with obvious white terminal band, which is unique among similar species. **Habitat** Pelagic, in Antarctic, subantarctic and temperate waters. **Voice** Silent at sea. Noisy at colonies, both in flight and from burrows. **Where to see** South of Antarctic Convergence, near Cape Horn and, occasionally, Beagle Channel (Magallanes Region). Sometimes follows fishing boats, but is generally not attracted to them.

Antarctic Prion
Pachyptila desolata R
L 28–29cm. **W** 57–61cm. Sexes identical. **ID** All prions are difficult to identify at sea. Unmistakable in hand, by broad bill, dark ear-coverts and thin white eyebrow. **Habitat** Pelagic, mostly in Antarctic and subantarctic waters. Also temperate, even tropical waters, in Humboldt Current. **Voice** Silent at sea. Vocal only at breeding sites, where gives various calls, but is rarely noisy. **Where to see** Antarctic and subantarctic waters, and on nesting islands (Magallanes Region). Sometimes attracted to fishing boats by chum.

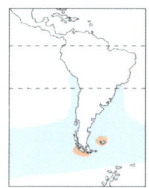

Slender-billed Prion
Pachyptila belcheri R
L 27cm. **W** 56cm. Sexes identical. **ID** All prions are difficult to identify at sea, but this is probably the most easily separated, by its thin bill. Unmistakable in the hand by its slender bill and extensive white face. **Habitat** Pelagic, in subantarctic and temperate waters; also tropical waters of Humboldt Current. **Voice** Silent at sea. Various calls heard at colonies, which can be very noisy at night. **Where to see** Subantarctic waters. Around Noir Island (Magallanes Region). Not attracted to fishing boats.

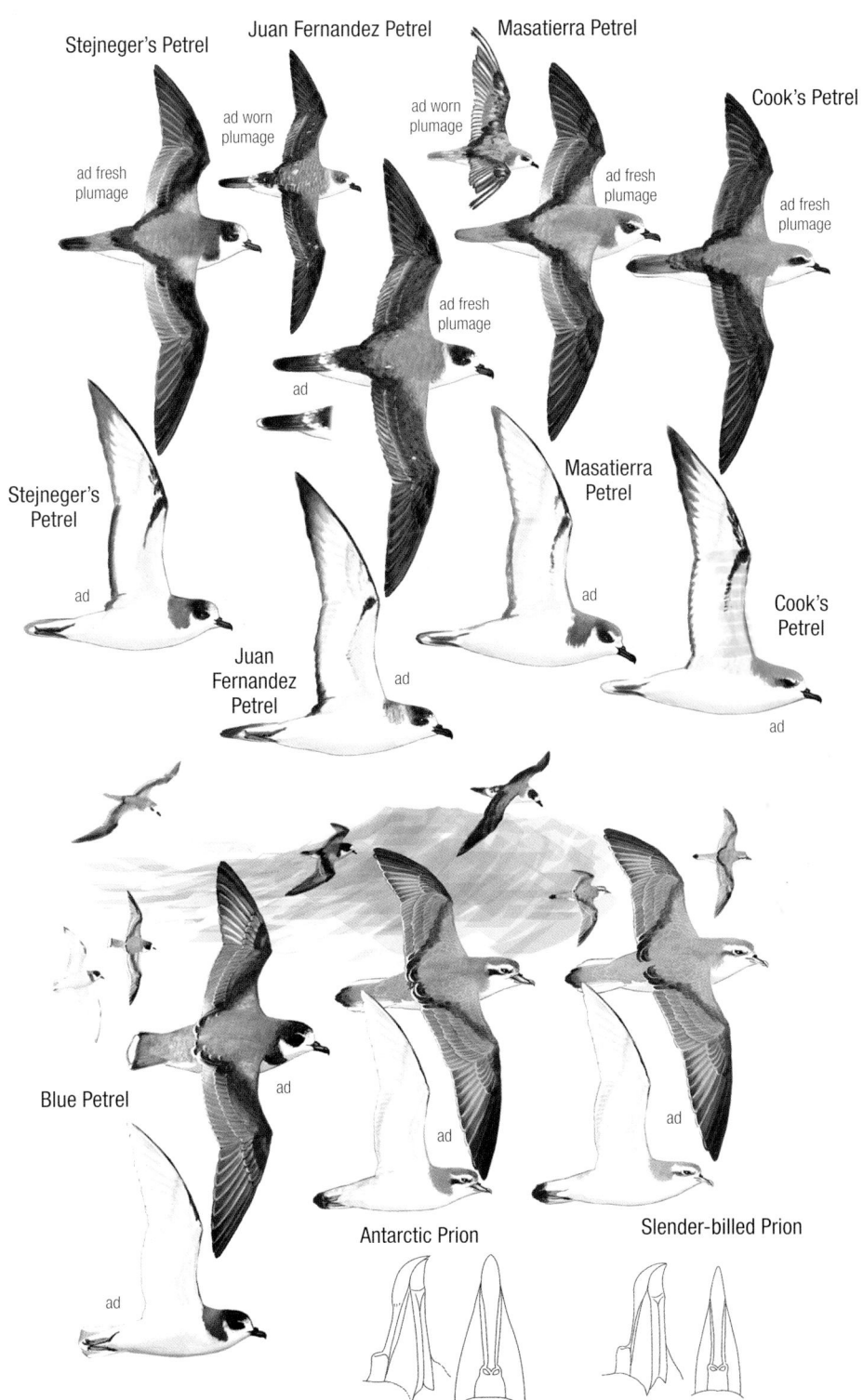

PLATE 11: PETRELS AND SHEARWATERS III

Kermadec Petrel
Pterodroma neglecta   R
L 37–40cm. W 91cm. Sexes identical. **ID** Very variable. Similar to Herald Petrel. Outer primaries have whitish shafts (dorsal view) that can appear like a patch at long range. **Habitat** Pelagic, mainly in temperate and tropical waters of Pacific. **Voice** Silent at sea, but noisy, both in flight and from the ground, at colonies. **Where to see** Breeds on Alejandro Selkirk Island, Juan Fernandez archipelago (Valparaíso Region). Also on San Félix and San Ambrosio (Desventuradas Islands), and Easter Island (Valparaíso Region). Not attracted to boats.

Herald Petrel
Pterodroma heraldica   R
L 36–37cm. W 90–97cm. Sexes identical. **ID** Similar to pale morph Kermadec Petrel and Murphy's Petrel. Dark upperwing, without whitish shafts to outer primaries (compared to Kermadec) and much white in underwing, plus square-ended tail, compared to Murphy's. **Habitat** Pelagic, mainly in temperate and tropical waters of Pacific. **Voice** Silent at sea, but very vocal at colonies, where gives a characteristic *Kekekekekekekekekekekeke....* **Where to see** Easter Island and Salas y Gómez Island (Valparaíso Region). Not attracted to boats.

Phoenix Petrel
Pterodroma alba   R
L 33–38cm. W 83cm. Sexes identical. **ID** Unmistakable. White underparts and black hood extending to throat. White undertail. **Habitat** Pelagic, in central Pacific, mainly in tropical waters. **Voice** Silent at sea, but very noisy at colonies. **Where to see** On Easter Island (Valparaíso Region). Not attracted to boats.

Wedge-tailed Shearwater
Ardenna pacifica   R
L 38–47cm. W 97–110cm. Sexes identical. **ID** Similar to Pink-footed (Plate 9) and Christmas Shearwaters. Two morphs and variable bill colour. Note long wedge-shaped tail and pink legs spotted dark. **Habitat** Pelagic, in Pacific and Indian Oceans, mainly in tropical waters, but also temperate seas. **Voice** Silent at sea, but vocal from burrows in colonies. **Where to see** Easter Island (Valparaíso Region). Not usually attracted to boats, but sometimes comes to chum.

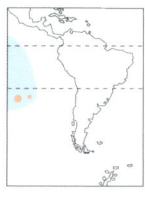

Murphy's Petrel
Pterodroma ultima   R
L 35–37cm. W 97cm. Sexes identical. **ID** Similar to Kermadec Petrel dark morph and Henderson Petrel, depending on light. All grey, with whitish chin and part of throat, and pale base to lower mandible. **Habitat** Pelagic, in eastern and central Pacific, mainly in tropical and, especially in Northern Hemisphere, temperate waters. **Voice** Silent at sea. Vocal when arriving at nest by night, where can be very noisy, giving shrill squawks in flight and on ground, similar to other petrels, but also has a characteristic, long and not very sharp hooting that initially accelerates before slowing somewhat (*Wowowowowowo...ho-ho-ho-hoho-ho...*). **Where to see** Easter Island and Salas y Gómez Island (Valparaíso Region) during breeding season (August–October). Not attracted to boats.

Christmas Shearwater
Puffinus nativitatis   R
L 33–36cm. W 83–90cm. Sexes identical. **ID** Similar to Sooty Shearwater (Plate 9). Dark chocolate-brown body and barely paler underwings. Dark legs. **Habitat** Pelagic, in central Pacific. Mainly in tropical waters, also temperate seas. **Voice** Silent at sea, but at colonies utters very plaintive-sounding calls (*Kaw kaw kaw kaw...*). **Where to see** Easter Island and Salas y Gómez Island (Valparaíso Region).

Henderson Petrel
Pterodroma atrata   M/R?
L 36–37cm. W 90–97cm. Sexes identical. **ID** Similar to Kermadec Petrel dark morph, which has a similar range. Dark below, without white in primaries. The secondaries can appear pale below, depending on the light, but there are no white shafts in the upperwing. **Habitat** Pelagic, in tropical waters of Pacific, sometimes ranging to temperate seas. **Voice** Silent at sea, but permanently noisy at colonies. **Where to see** Easter Island (Valparaíso Region). Not attracted to boats.

Subantarctic Shearwater
Puffinus elegans   V
L 28cm. W 53–62cm. Sexes identical. **ID** Recently split. Recalls a diving-petrel in coloration, but flight like other shearwaters. Small, with a compact body and fast fluttering wingbeats. **Habitat** Pelagic, in temperate waters. **Voice** Silent at sea. **Where to see** Most likely in extreme south and near Chiloé Island (Los Lagos Region). **Note** Formally treated as conspecific with Little Shearwater *P. assimilis*.

PLATE 12: PETRELS AND SHEARWATERS IV

Atlantic Petrel
Pterodroma incerta   V
L 43–44cm. **W** 104cm. Sexes identical. **ID** Unmistakable. White breast and belly, contrasting with rest of dark brown body, head and neck. Compared to Phoenix Petrel, note dark undertail-coverts (white in Phoenix). **Habitat** Pelagic, in temperate waters of Atlantic. **Voice** No information. **Where to see** Most likely in Drake Passage; south of Cape Horn (Magallanes Region). Does not follow ships, but occasionally approaches them.

Manx Shearwater
Puffinus puffinus   V
L 31–35cm. **W** 75–84cm. Sexes identical. **ID** Unmistakable in Chile. Slim, typical *Puffinus* body, dorsally black and ventrally white. **Habitat** Pelagic, in Atlantic Ocean, especially temperate and cold waters. **Voice** Silent at sea, but noisy at colonies. **Where to see** Scattered records from Humboldt Current. Does not follow ships and not attracted to chum.

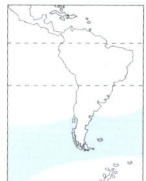

Grey Petrel
Procellaria cinerea   M
L 48–51cm. **W** 117–125cm. Sexes identical. **ID** Unmistakable, by yellowish bill and grey upperparts and head. **Habitat** Pelagic, in subantarctic and temperate waters. Visits tropical waters following Humboldt Current, perhaps as far as Peru. **Voice** Silent at sea, but loud calls heard at colonies. **Where to see** Most likely in Drake Passage; south of Cape Horn (Magallanes Region). Follows boats and is attracted to fish waste.

White-headed Petrel
Pterodroma lessonii   M
L 43–46cm. **W** 99–107cm. Sexes identical. **ID** Unmistakable. All-white head, except black patch around eye. **Habitat** Pelagic, mainly in Antarctic and subantarctic waters, but also cold temperate seas. **Voice** Silent at sea. At colonies, vocalises in flight and to lesser extent from the ground. **Where to see** Drake Passage, subantarctic and Antarctic waters to edge of pack ice. Does not follow fishing boats, but will briefly approach them.

Mottled Petrel
Pterodroma inexpectata   V/M?
L 32–36cm. **W** 84–92cm. Sexes identical. **ID** Unmistakable, white-and-black underwing pattern. **Habitat** Pelagic, in Pacific, mostly in cold waters (Antarctic and subantarctic seas, as well as Bering Sea in the north), but also temperate and tropical waters. **Voice** Silent at sea. Very noisy at colonies, both in flight and on the ground. **Where to see** Drake Passage. Does not follow fishing boats.

Kerguelen Petrel
Aphrodroma brevirostris   V
L 33–36cm. **W** 80–82cm. Sexes identical. **ID** Unmistakable, although plumage colour varies significantly with light, causing the head to look darker than the body. **Habitat** Pelagic, mainly in subantarctic and even Antarctic waters. **Voice** Silent at sea. At colonies very noisy, both in flight and from the ground. **Where to see** Most likely in Drake Passage; south of Cape Horn (Magallanes Region). Does not follow fishing boats, but is occasionally attracted to fish waste.

Snow Petrel
Pagodroma nivea   R
L 34–36cm. **W** 80cm. Sexes identical. **ID** Unmistakable. Angelic white petrel, with a stocky body, fan-shaped tail and black bare parts. **Habitat** Antarctic waters, where usually closely linked to ice. **Voice** Silent at sea, but very noisy at colonies. **Where to see** Antarctic waters, mainly in areas with pack ice. Normally does not follow boats.

Antarctic Petrel
Thalassoica antarctica   R
L 43cm. **W** 104cm. Sexes identical. **ID** Unmistakable, brown-and-white pattern on body. **Habitat** Pelagic and circumpolar in Southern Hemisphere, in Antarctic and subantarctic waters. Usually near floating ice and Antarctic coasts, but will wander to cold (temperate) seas. **Voice** Silent at sea, but very vocal at colonies, where calls are cackles, interspersed by guttural sounds. **Where to see** Within Antarctic Convergence and especially east of Antarctic Peninsula. Usually does not follow fishing boats, and is not attracted to chum.

PLATE 13: STORM-PETRELS I

Markham's Storm-Petrel
Oceanodroma markhami **R**
L 24cm. **W** 56cm. Sexes identical. **ID** Unmistakable in Chile. A large, wholly dark (blackish) and relatively slow-flying storm-petrel. **Habitat** Pelagic, in tropical, and marginally temperate, waters of eastern Pacific. **Voice** Silent at sea, but vocalises softly in flight when arriving at colonies, and also from inside nests. **Where to see** Humboldt Current, off Arica (Arica y Parinacota Region), Iquique (Tarapacá Region) or Antofagasta (Antofagasta Region).

Ringed Storm-Petrel
Oceanodroma hornbyi **R**
L 21–29cm. Sexes identical. **ID** Unmistakable. White underparts with a thin but well-defined dark collar, black cap reaching to the eyes, and forked tail. **Habitat** Eastern Pacific, usually in waters 50–160km offshore, mainly in tropical and also temperate (but warm subtropical) waters of Humboldt Current. Recently discovered nesting in the Atacama Desert. **Voice** Silent at sea. **Where to see** Off Antofagasta or Mejillones (Antofagasta Region). [Alt. Hornby's Storm-Petrel]

White-faced Storm-Petrel
Pelagodroma marina **M**
L 20cm. **W** 42cm. Sexes identical. **ID** Similar to Ringed and Polynesian Storm-Petrels. Long black patch around eye, and dark tail. **Habitat** Pelagic, in tropical and temperate waters. **Voice** Silent at sea, but noisy at colonies, vocalising from the ground and in flight. **Where to see** Juan Fernández archipelago and environs (Valparaíso Region).

White-vented Storm-Petrel
Oceanites gracilis **R**
L 14–18cm. Sexes identical. **ID** Similar to Wilson's Storm-Petrel. Rump and belly (also part of breast) white, and also note short (sometimes hardly evident) black longitudinal line connecting flanks and thighs. **Habitat** Pelagic, in Pacific Ocean, but mainly associated with Humboldt Current. Mainly tropical waters, but also temperate (warmer) waters. **Voice** Chicks give gentle chirps while in the nest. Other calls unknown to us. **Where to see** Off Arica (Arica y Parinacota Region), or around Chungúngo Island (Coquimbo Region), where it nests. [Alt. Elliot's Storm-Petrel]

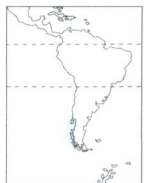

Pincoya Storm-Petrel
Oceanites pincoyae **Re?**
L c.16cm. **W** 32–33cm. Sexes identical. **ID** Similar to Wilson's Storm-Petrel. White below not extensive. In dorsal view, the greater and inner secondaries are edged white to pale grey; in ventral view, the inner coverts are white. Combined, they give the impression of a whitish appearance to the wings. **Habitat** Southern Chile's inshore waters. No other information. **Voice** No information. **Where to see** Known only from the Reloncaví Sound and near Chiloé Island (Los Lagos Region). **Note** Nesting has yet to be found.

Wilson's Storm-Petrel
Oceanites oceanicus **R**
L 18cm. **W** 41cm. Sexes identical. **ID** Similar to White-vented, Pincoya, Wedge-rumped and Black-bellied Storm-Petrels. Note white rump, square tail and variable white on underparts, which declines to almost none in Antarctic *O. o. exasperatus*, and just reaches tarsus or slightly in front of it in *O. o. chilensis* (Humboldt Current and southern Chile; Fuegian Storm-Petrel). No whitish in underwing of *O. o. exasperatus*, but *O. o. chilensis* has whitish line on underwing-coverts. **Habitat** Antarctica (*O. o. exasperatus*) or mainly linked to Humboldt Current (*O. o. chilensis*). **Voice** Silent at sea, but sometimes gives soft calls when foraging; at colonies is vocal, mainly from the ground, but not especially noisy. **Where to see** Antarctic waters and the Humboldt Current.

Wedge-rumped Storm-Petrel
Oceanodroma tethys **R**
L 17cm. Sexes identical. **ID** Similar to Wilson's Storm-Petrel, but has extensive white at base of tail (rump and uppertail-coverts). **Habitat** Pelagic, in tropical and subtropical waters of eastern Pacific. **Voice** Silent at sea, but vocalises frequently at colonies, although it is not noisy. **Where to see** In spring around Grande Island (Atacama Region), where it nests. Also Humboldt Current, between Caldera (Atacama Region) and Arica (Arica y Parinacota Region).

PLATE 14: STORM-PETRELS II AND DIVING PETRELS

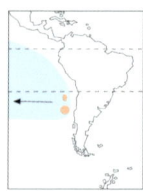

White-bellied Storm-Petrel
Fregetta grallaria   R
L 18–22cm. **W** 44–48cm. Sexes identical. **ID** Similar to Black-bellied Storm-Petrel. Lower breast, belly and undertail-coverts white. Well-defined white area on underwing. Dark morph not recorded in Chile. When in fresh plumage, upperparts have fine white edges, which are lost with wear. **Habitat** Pelagic, in temperate and tropical waters. **Voice** Silent at sea, but gives frequent soft calls at colonies, most commonly a repetitive *Pew-pew-pew-pew-pew-pew*..... **Where to see** Juan Fernández archipelago, and San Félix and San Ambrosio (Desventuradas Islands) (Valparaíso Region).

Black-bellied Storm-Petrel
Fregetta tropica   M
L 20cm. **W** 46–48cm. Sexes identical. **ID** Similar to White-bellied and Wilson's Storm-Petrels. White breast and belly with a black longitudinal midline, which is not always visible (due to angle or its variability), and black undertail-coverts. Whitish underwing not well defined, more diffuse. In fresh plumage, upperparts have fine white edges, which are lost with wear. **Habitat** Pelagic, mainly Antarctic and subantarctic waters, more rarely temperate seas and marginally in tropical waters. **Voice** Silent at sea. At colonies both adults and chicks are very vocal. Most commonly heard voice is that of adult, a sad whistle, repeated constantly. **Where to see** Drake Passage.

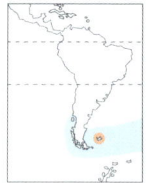

Grey-backed Storm-Petrel
Garrodia nereis   M
L 18cm. **W** 39cm. Sexes identical. **ID** Unmistakable. Grey back, rump and tail, the latter paler with a black terminal band. Underwing almost all white. **Habitat** Pelagic, in temperate (mainly), subantarctic and Antarctic waters. **Voice** Silent at sea, but vocalises from the ground at colonies. **Where to see** Southern waters, around Cape Horn (Magallanes Region) and Drake Passage.

Polynesian Storm-Petrel
Nesofregetta fuliginosa   R
L 21–25cm. **W** 51cm. Sexes identical. **ID** Similar to Ringed Storm-Petrel. A large storm-petrel, with black head and chest, divided by white throat, which projects like a partial collar. **Habitat** Pelagic in tropical waters of central Pacific. **Voice** No information. **Where to see** Around Salas y Gómez Island (Valparaíso Region), where it has nested.

Peruvian Diving Petrel
Pelecanoides garnoti   R
L 22–25cm. Sexes identical. **ID** Unmistakable in range. No obvious contrast between upper- and underparts, and has a robust bill. **Habitat** Waters just beyond the surf, in a narrow strip of a few km. **Voice** Silent at sea, but gives quiet calls, vaguely recalling a dove, at colonies. **Where to see** Humboldt Current, from Valparaíso north; near Chungúngo, Choros and Damas islets (Coquimbo Region); off Arica (Arica y Parinacota Region).

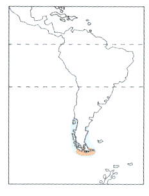

Magellanic Diving Petrel
Pelecanoides magellani   R
L 19–21cm. Sexes identical. **ID** Similar to Common Diving Petrel. White of underparts penetrates towards the hindneck, around the ear-coverts. The inner wing is whitish. In fresh plumage the upperparts have pale grey fringes to the feathers. **Habitat** Inshore waters, including channels and straits. **Voice** No information. **Where to see** Chacao Channel (Los Lagos Region); Magellan Strait (Magallanes Region).

Common Diving Petrel
Pelecanoides urinatrix   R
L 20–23cm. Sexes identical. **ID** Similar to Magellanic Diving Petrel. Boundary between dark upperparts and white underparts diffuse, especially on neck and head. Inner wing is grey. **Habitat** Inshore waters, in open sea but also channels around coasts. **Voice** No information. **Where to see** Fjords and channels around Aysén (Aysén Region).

PLATE 15: CORMORANTS

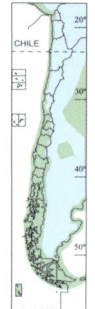

Neotropic Cormorant
Phalacrocorax brasilianus   R
L 71–76cm. **W** 102cm. Sexes identical. **N** <4,600m. **C** <3,000m. **S&A** <500m. **ID** Adult unmistakable. Juvenile has dark brown body, paler below, and sometimes with whitish throat. **Habitat** Marine coasts, near the shoreline. Also rivers, lakes and inland lagoons. **Voice** Silent at sea. At colonies, noisy, giving guttural and rough sounds, like the snorts of a pig. **Where to see** Widespread in Chile. Arica port and Chungará Lake (Arica y Parinacota Region), Valparaíso coast (Valparaíso Region), or any port or cove along the coast.

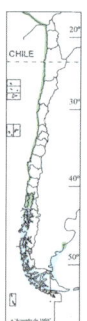

Red-legged Cormorant
Phalacrocorax gaimardi   R
L 60–71cm. **W** 89cm. Sexes identical. **ID** Unmistakable. Unique, grey-bodied, red-legged cormorant. **Habitat** Marine coasts and nearby islets. Large rocks and cliffs. **Voice** Silent at sea. Colonies are not noisy, the birds giving mostly sharp calls. **Where to see** Cliffs south of Arica, Alacrán Peninsula (Arica y Parinacota Region), Pisagua (Antofagasta Region), Cobquecura (Bio-Bio Region), Boca Budi (Araucanía Region), Chacao Channel, Puñihuil (Los Lagos Region).

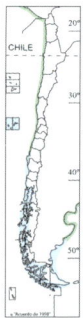

Guanay Cormorant
Phalacrocorax bougainvilliiorum   R
L 68–76cm. Sexes identical. **ID** Unmistakable, with white throat and abdomen. **Habitat** Feeds in pelagic waters, returning daily to coasts, nearby islets, large rocks and cliffs. **Voice** Silent at sea, but colonies are noisy. **Where to see** Cliffs south of Arica, Alacrán Peninsula (Arica y Parinacota Region), Pájaro II islet, Choros Island (Coquimbo Region), San Antonio port (Valparaíso Region), Pupuya (O'Higgins Region).

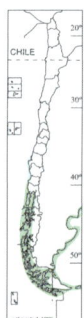

Magellanic Cormorant
Phalacrocorax magellanicus   R
L 62–66cm. Sexes identical. **ID** Unmistakable, once the red cere and bare facial skin, and white spot on the ear-coverts, are seen. **Habitat** Coasts and nearby islets. Large rocks and cliffs. **Voice** Silent at sea. At breeding sites gives low calls, which sound midway between a trumpet and a grunting pig. **Where to see** Puñihuil (Los Lagos Region), road to Fuerte Bulnes, road to Chilota Bay, Porvenir (Magallanes Region). [Alt. Rock Shag]

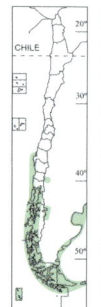

Imperial Cormorant
Phalacrocorax atriceps   R
L 75–78cm. Sexes identical. **ID** Almost indistinguishable from Antarctic Cormorant. Two forms, which differ in the way black-and-white is divided on the cheek. **Habitat** Coasts, channels and large freshwater lakes. **Voice** Silent at sea. At breeding sites gives low calls, which sound midway between a trumpet and a grunting pig. **Where to see** Pelluco, Chacao Channel (Los Lagos Region), Puerto Natales Bay, road to Fuerte Bulnes, road to Chilota Bay, Porvenir (Magallanes Region). **Note** Of the two forms, the commonest in the Pacific has a curved line separating black colour from white cheeks, while the commonest in the Atlantic has a straight line separating black colour from white cheeks. They form mixed colonies in the far south, but are sometimes treated as subspecies, *atriceps* (Pacific) and *albiventer* (Atlantic). [Alt. Imperial Shag]

Antarctic Cormorant
Phalacrocorax bransfieldensis   R
L 75cm. Sexes identical. **ID** Unmistakable in range, but elsewhere would be almost indistinguishable from Imperial Cormorant. **Habitat** Antarctic coasts and islets. **Voice** Silent at sea. Colonies, on the other hand, can be very noisy. **Where to see** Antarctic Peninsula (Magallanes Region).
Note Antarctic Cormorant is sometimes treated as conspecific with Imperial Cormorant.

PLATE 16: PELICANS AND FRIGATEBIRD

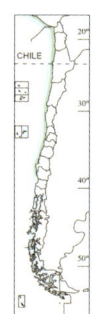

Peruvian Pelican
Pelecanus thagus R
L 137–152cm. **W** 228cm. Sexes identical. **ID** Similar to Brown Pelican, but has whitish wing-coverts strongly contrasting with secondaries. Also, bluish-white gular pouch. Larger than Brown Pelican. **Habitat** Coasts to a few km offshore. **Voice** Lacks muscles in the syrinx. Silent at sea, but at breeding sites emits low-pitched sounds, without ever being noisy. **Where to see** Any port, cove or coastal rock between Arica and Chiloé Island, less regularly even further south.

Great Frigatebird
Fregata minor R
L 93–102cm. **W** 213–218cm. Sexually dimorphic. **ID** Unmistakable in Chile. Very large, rakish seabird, with W-shaped wings, deep wingbeats and long, narrow bill; note clear sexual dimorphism and age-related variation. **Habitat** Pelagic, in tropical waters. **Voice** Silent at sea, but around nests may call both from the ground and in flight. **Where to see** Easter Island (Valparaíso Region).

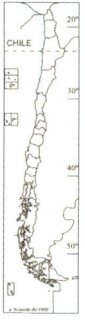

Brown Pelican
Pelecanus occidentalis V
L 117–132cm. **W** 203cm. Sexes identical. **ID** Differs from Peruvian Pelican by its dark gular pouch and upperwings (secondaries not obviously contrasting with coverts). **Habitat** Coasts to a few km offshore. **Voice** Lacks muscles in the syrinx. Silent at sea, but at breeding sites emits low-pitched sounds, without ever being noisy. **Where to see** Vagrant to Arica (Arica y Parinacota Region) and Iquique (Tarapacá Region). Appears to be expanding south in South America.

PLATE 17: BOOBIES AND TROPICBIRDS

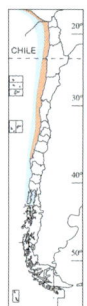

Peruvian Booby
Sula variegata   R
L 73cm. **W** 138cm. Sexes identical. **ID** Adult unmistakable. Juvenile similar to Blue-footed Booby (juvenile and immature). Grey belly and chest, plus greyish neck and head. Fine whitish lines on mantle and wing-coverts at all ages. **Habitat** Always tied to the Humboldt Current. Rare offshore. **Voice** Silent at sea, but at colonies calls of male and female are recognisably different, with the male sounding sharper, and female lower and a little like a trumpet. **Where to see** Widespread. Cliffs south of Arica (Arica y Parinacota Region); Isla Pájaros (Atacama Region), cliffs of Tirilluca (Valparaíso Region), etc.

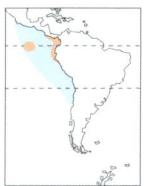

Blue-footed Booby
Sula nebouxii   R
L 81–85cm. **W** 158cm. Sexes identical. **ID** Adult unmistakable. Juvenile and immature could be confused with juvenile Peruvian Booby, but have white belly and brownish-grey head to breast. At all ages dark brown wing-coverts lack whitish edges. **Habitat** Coasts and oceanic islands, in tropical waters of Pacific. **Voice** Silent at sea, but vocal at colonies. **Where to see** Vagrant to mainland, where probable around Arica (Arica y Parinacota Region); has nested on San Félix and San Ambrosio Islands (Valparaíso Region).

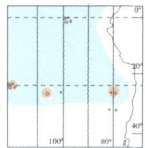

Masked Booby
Sula dactylatra   R
L 75–81cm. **W** 150–158cm. Sexes identical. **ID** Adult unmistakable. Juvenile shows strong contrast between white underparts and brown head and neck. **Habitat** Pelagic, mostly in tropical waters. **Voice** Silent at sea, but may give a high-pitched call prior to plunge-dives. At colonies noisy, with males (more like a whistle) and females (lower, more croaking) audibly different. **Where to see** Easter Island; and San Félix and San Ambrosio Islands (Valparaíso Region).

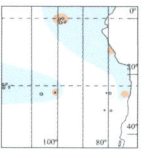

Red-billed Tropicbird
Phaethon aethereus   R
L 61cm. **W** 110cm. **Tail** 100cm. Sexes identical. **ID** Adult unmistakable. Juvenile has strong black-tipped yellow bill and black mask that joins on hindneck. **Habitat** Pelagic, mainly in tropical waters. Some records in temperate seas. **Voice** Vocal in flight and on ground at colonies. Calls usually strident, a repetitive high-pitched *Kek-kek-kek-kek-kek-kek-kek-kek*.... **Where to see** Around Salas y Gómez Island (Valparaíso Region).

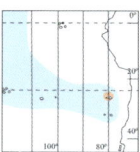

White-tailed Tropicbird
Phaethon lepturus   R
L 41cm. **W** 93cm. **Tail** 80cm. Sexes identical. **ID** Adult unmistakable. Juvenile similar to juvenile Red-billed, but has a pale yellowish bill with a black tip and the black mask does not join on the hindneck. **Habitat** Pelagic, mainly in tropical waters, occasionally temperate seas. **Voice** Silent at sea. At colonies gives diverse calls, but is the least noisy of the three tropicbirds. Calls in flight, alone and when in pairs or groups. **Where to see** Easter Island and San Felix and San Ambrosio Islands (Valparaíso Region).

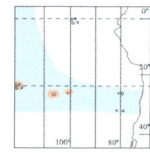

Red-tailed Tropicbird
Phaethon rubricauda   R
L 46cm. **W** 107cm. **Tail** 91cm. Sexes identical. **ID** Adult unmistakable. Juvenile has dirty greyish-yellow bill, and no mask. **Habitat** Pelagic, mainly in tropical waters. **Voice** Silent at sea. At colonies gives loud calls in air or when defending the nest. **Where to see** Easter Island, San Félix and San Ambrosio Islands (Valparaíso Region) and Salas y Gómez Islands (Valparaíso Region).

PLATE 18: JAEGERS AND SKUAS

JAEGERS (visitors). More identifiable by structure and behaviour than coloration. All have at least two well-defined morphs (light and dark) and an extensive range of intermediates. A sample of adult and juvenile plumages is shown, all of which can be observed in Chilean waters.

SKUAS (mainly resident). Difficult to separate at sea (except Chilean Skua). Robust birds, with heavy flight, and characteristic white spot at base of primaries. Hybridisation has been recorded among all three species.

Long-tailed Jaeger
Stercorarius longicaudus   **M**
L 26–31cm. Sexes identical. **ID** The smallest and slightest of the jaegers. Adult is slim-bodied, with a short bill, narrow wings (two white shafts in upperwing and no whitish in underwing) and a relatively long tail. Birds in first year have whitish underwing primaries. Flight distinctively weaves up and down. Silhouette and flight resemble a tern. **Habitat** Pelagic in Chile, where apparently tied to Humboldt Current. **Voice** Silent off Chile; vocal mostly when nesting. **Where to see** Humboldt Current, mainly off northern Chile, sometimes from coasts. [Alt. Long-tailed Skua]

Parasitic Jaeger
Stercorarius parasiticus   **M**
L 41–50cm. Sexes identical. **ID** Similar to other jaegers. Medium-sized, appearing better-proportioned than others. Adult has long, thin bill, upperwing shows four to six white shafts, and flight is usually direct. **Habitat** In Chile is tied to coasts, rather than being pelagic. **Voice** Silent at sea, but may utter high-pitched squawks, reminiscent of a Kelp Gull (*Qweea... qweea ...*), when competing for food. **Where to see** Only in Humboldt Current, mostly off northern Chile. Kleptoparasitises terns. [Alt. Arctic Skua]

Pomarine Jaeger
Stercorarius pomarinus   **M**
L 48–58cm. Sexes identical. **ID** Similar to other jaegers, but larger. Heavy-bodied, with a broad chest and broad wings (especially at the base). Adult shows no white in underwing-coverts, and has a relatively large bill. Slower wingbeats than other jaegers. **Habitat** Pelagic in Chile, although may approach the coast. **Voice** Silent at sea, except sharp squawks (*Keew ... Keew ...*) at food. Most vocal when breeding. **Where to see** Humboldt Current, especially off northern Chile. [Alt. Pomarine Skua]

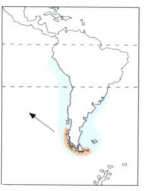

Chilean Skua
Stercorarius chilensis   **R**
L 55–61cm. Sexes identical. **ID** Unmistakable. Cinnamon-brown or rufous plumage, and dark crown with capped appearance, diagnostic at all stages. Bluish-grey bill with dark tip unique among resident skuas. **Habitat** Coastal and pelagic. Tied to coasts, although can venture far out to sea. Fjords, southern channels, even inland lakes. **Voice** Generally silent, but in territorial displays calls while extending the wings and neck upwards. Monosyllabic, recalling some ducks, with a pronounced nasal tone (*Haek ... haek ...*), sometimes repeated several times with a variable tone, which can be very sharp towards the end. **Where to see** Humboldt Current, Magellan Strait, Puerto Natales Bay, Porvenir Bay, Beagle Channel (Magallanes Region).

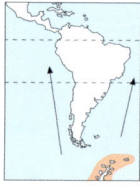

South Polar Skua
Stercorarius maccormicki   **R**
L 53–55cm. Sexes identical. **ID** Several morphs. Pale morph unmistakable. Dark morph and intermediates can be confused with Brown Skua. Underwing has dark lesser coverts, pale median coverts and dark greater coverts, forming a contrasting pattern. Short, somewhat thin bill (smallest among resident skuas). Juvenile has coverts with paler rachis. **Habitat** Pelagic and coastal. When breeding closely tied to coasts and interior of Antarctica. Post-breeding, migrates to Northern Hemisphere and is pelagic. **Voice** Vocal mainly in territory defence or alarm. Among calls, gives a raspy *Aahk-aahk-aahk-aahk ...*, and a choked laugh (*Ah-ah-ah-ah-ah-ah-ah...*). **Where to see** Antarctica (Magallanes Region). Moves north in autumn, when pelagic. Some partially follow Humboldt Current.

Brown Skua
Stercorarius antarcticus   **R**
L 53–64cm. Sexes identical. **ID** Similar to South Polar Skua, but has uniformly dark inner wing-coverts, and grey secondary-coverts without any obvious pattern. Head and neck finely striated ochre. Massive bill, but obvious only at close range. The largest of this group. **Habitat** Marine, tied to coasts. Juveniles pelagic. **Voice** Rather silent, vocal mainly during territorial displays with raised wings, or in aggression, sometimes giving low sounds when in groups. Higher-pitched than other species in Chile. **Where to see** Antarctic Peninsula and nearby islands (Magallanes Region).

PLATE 19: GULLS I AND SHEATHBILL

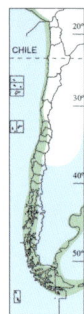

Kelp Gull
Larus dominicanus **R**
L 58cm. Sexes identical. 0–800m. **ID** Similar to Belcher's Gull. Juvenile has all-black bill; immature whitish/yellowish with black tip; and adult yellow with a red gonys. Juvenile has black tail; second-year bird has tail with a variable amount of black, like a band, over white. Adult has pale iris. **Habitat** Coasts and a few km offshore. Ventures inland along rivers. Also agricultural areas and cities. **Voice** Very vocal. Common name on Chiloé is onomatopoeic based on the commonly heard *kowk-kowk-kowk-kowk-kowk...*, which is always loud. Also *yap-yap-yap-yap....* **Where to see** Ports, beaches, rubbish dumps. Coasts throughout central and southern Chile. Also Antarctica.

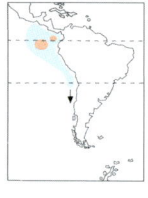

Swallow-tailed Gull
Creagrus furcatus **V**
L 51–57cm. Sexes identical. **ID** Similar to Sabine's Gull. Obviously large eyes at all ages. Grey-and-white wings, with five outer primaries tipped and edged black, but the rest of the vane white. Forked tail. **Habitat** Pelagic, inshore waters off South America and Galápagos Islands, but recorded up to 500km out to sea. **Voice** Usually silent at sea, but calls occasionally in flight, and is frequently heard around its colonies, where it gives a mix of very sharp and throaty or raspy notes, some recalling a pig. **Where to see** Humboldt Current, especially in the north, off Arica (Arica y Parinacota Region), Iquique (Tarapacá Region) and Antofagasta (Antofagasta Region).

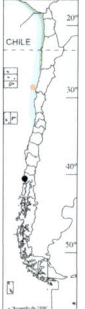

Belcher's Gull
Larus belcheri **R**
L 48–51cm. Sexes identical. 0–100m. **ID** Similar to Kelp Gull. Yellow legs and yellow bill, with a red-and-black tip. Broad black tail-band at all ages. Adult has dark iris. **Habitat** Coasts and a few km offshore. Prefers rocky beaches. **Voice** Vocalises frequently, but is not especially noisy. Calls resemble those of Kelp Gull, usually a repeated *kaoww-kaow-kaow-kaow-kaow-kaow....* **Where to see** Ports and beaches, around Arica and Alacrán Peninsula (Arica y Parinacota Region), Iquique (Tarapacá Region) and Caldera (Atacama Region).

Snowy Sheathbill
Chionis albus **R**
L 34–41cm. Sexes identical. 0–300m. **ID** Unmistakable. Dumpy white body, vaguely recalling a dove. **Habitat** Coasts. Around colonies of birds and marine mammals. **Voice** Generally silent, but gives hoarse, harsh and low-pitched calls, which stand out among the bustle of penguin colonies. Most frequently heard is a short, very hoarse, rough, repeated *aaghk-aaghk-aahgk-aaghk-aaghk....* **Where to see** Antarctic Peninsula and nearby islets. Accidental in winter on continent.

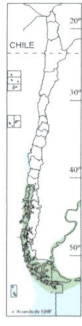

Dolphin Gull
Leucophaeus scoresbii **R**
L 40–46cm. Sexes identical. 0–50m. **ID** Unmistakable. Juvenile has all-black bill; immature horn/pinkish with black tip. Heavy red bill and red legs in adult. **Habitat** Rocky coasts, often near settlements. **Voice** Vocalises frequently. Very sharp calls, almost strident, compared to other species in same areas. **Where to see** Streams and rubbish dumps, around Punta Arenas, Puerto Natales Bay, Porvenir Bay (Magallanes Region).

Grey Gull
Leucophaeus modestus **R**
L 44–46cm. Sexes identical. 0–1,400m. **ID** Unmistakable. Juvenile has brown body, sometimes with cinnamon hue. Adult has all-grey body, but head white when breeding. **Habitat** Sandy coastal beaches and a few km offshore. Nests almost exclusively in Atacama Desert, up to 100km inland. **Voice** Very noisy. Typical is a characteristic, very sharp *Keeeaaaaaaa ... Keeeaaaaaa...*, with a vibrato towards end, given both at rest and in flight. **Where to see** Sandy beaches between Arica (Arica y Parinacota Region) and central Chile, but mainly in the north during breeding season. **Note** In 2015, a colony was found at Mejillones (Antofagasta Region) with up to 140 young, the first and only known coastal colony of this species.

PLATE 20: GULLS II

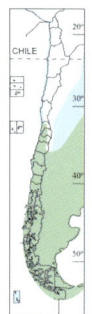

Brown-hooded Gull
Chroicocephalus maculipennis R
L 35–38cm. Sexes identical. 0–1,000m. **ID** Similar to Franklin's and Andean Gulls. Juvenile and immature have pale bill (orange-pinkish) with black tip; adult has four white outer primaries (when fresh with black tips, but these quickly wear off). **Habitat** Coasts, rivers, lakes, lagoons with reeds (where it nests) and inland estuaries. Also agricultural areas and cities. **Voice** Very noisy. Varied calls, but most commonly heard is an alarmed *krrrreep-krrreep...*, given several times, and a more aggressive and rasping *Krreeeeeeeaaaaa ... Krreeeeeeeaaaaa....* **Where to see** El Peral lagoon, Rio Maipo wetland (Valparaíso Region), Lampa/Batuco system (Metropolitan Region), Torca lagoon (Maule Region), Budi lake (Araucanía Region), Chiloé National Park, Chacao Channel (Los Lagos Region), etc.

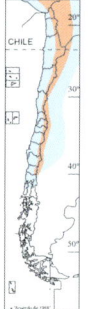

Andean Gull
Chroicocephalus serranus R
L 45–48cm. Sexes identical. **N** 0–4,600m. **C** & **S** 0–3,000m. **ID** Similar to Brown-hooded and Franklin's Gulls. Bill thin and smaller than other gulls. Juvenile and immature have blackish bill and legs. Adult has very dark red to black bill and legs. At all ages shows a white 'triangle' within the black wingtip. **Habitat** Wet areas, Andean bogs, salt flats, watercourses and lakes, especially in highlands, to more than 4,000m in the north, but lower in central and southern Chile. Also on coasts, less regularly, and mainly immatures. **Voice** Very noisy, repeatedly uttering sharp calls, especially around nest. Most commonly heard is an alarm call, repeated many times at variable intervals (*Kep ... Kep ... Kep ...*), and a very high-pitched, 'desperate'-sounding *Keeu..Keeu..Keeeeuu.. Keeeeeeuu...*, with each note drawn-out, which is most commonly given in flight. **Where to see** Lluta wetland, Lauca National Park, road to Surire (Arica y Parinacota Region), Atacama Salt Flat (Antofagasta Region), Colchane, Huasco Salt Flat (Tarapacá Region), Maule lagoon (Maule Region).

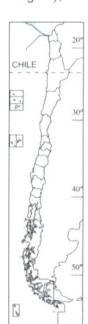

Grey-hooded Gull
Chroicocephalus cirrocephalus V
L 40–43cm. Sexes identical. 0–100m. **ID** Adult unmistakable. Immature like Brown-hooded and Andean Gulls, but note pale/pinkish bill (with dark tip) and legs; also wing pattern. Adult has grey hood (when breeding) and white iris (at all seasons). **Habitat** Coasts, but outside Chile also on inland lakes and swamps. **Voice** Not especially noisy, except at colonies. Gives a hoarse, scratchy *kahaaaa ... .kahaaaaa ... .kahaauuu ... kahaauuu...* that recalls Kelp Gull. **Where to see** Lluta wetland and Arica (Arica y Parinacota Region).

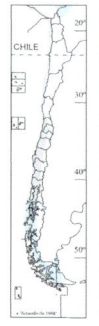

Franklin's Gull
Leucophaeus pipixcan M
L 32–38cm. Sexes identical. 0–800m. **ID** Similar to Brown-hooded, Andean and Laughing Gulls. Juvenile not seen in Chile, only first-winter (and adult). Dark grey mantle and black outer primaries with white tips. Broad white eye-crescents, but has short thin bill (compared to Laughing Gull). **Habitat** Coasts, especially sandy beaches with nearby fresh water. Via rivers, reaches up to more than 100km inland, where also occurs on lagoons. **Voice** Very noisy. Very sharp calls recall a laugh, comprising a single syllable repeated over and over, and sometimes drawn-out (*kwa-kwa-kwa-kwa-kwa-kwa-kwa-kwa ... kwuaa ... kwuaa ...*), as well as a very sharp *Peeeeep ... Peeeeep ... Peeeeep... Peeeeep... Peeeeep....***Where to see** Lluta wetland (Arica y Parinacota Region), El Peral lagoon, Rio Maipo wetland (Valparaíso Region), Concepción coast (Maule Region), Chacao Channel (Los Lagos Region).

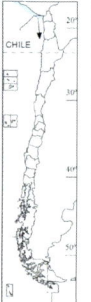

Laughing Gull
Leucophaeus atricilla V
L 38–46cm. Sexes identical. **ID** Very similar to smaller (and much commoner) Franklin's Gull, with longer and stronger bill. In flight note all-black outer primaries. **Habitat** Coasts. **Voice** A noisy bird; the most regularly heard voice vaguely recalls a laugh, and that of Grey Gull. **Where to see** Arica coast (Arica y Parinacota Region).

Sabine's Gull
Xema sabini V
L 27–34cm. Sexes identical. **ID** Similar to the larger Swallow-tailed Gull. At all ages note tricoloured wings: grey (or grey streaked brownish in juvenile) and white, with five outermost primaries black or blackish. Appears light in flight, recalling a tern. **Habitat** Pelagic in South America, although there are occasional records on coasts of Chile. **Voices**: Silent at sea, but very vocal on the breeding grounds, with somewhat tern-like calls. **Where to see** Humboldt Current, especially in northern Chile.

54

PLATE 21: TERNS I AND SKIMMER

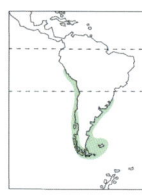

South American Tern
Sterna hirundinacea   R
L 40–43cm. Sexes identical. **ID** Similar to other *Sterna* terns. Large, slightly curved bill with a well-marked notch in the gonys in immature and adult. Juvenile and immature can show dark feathers in wing-coverts. **Habitat** Coasts and up to a few km offshore. Occasional at inland waters. **Voice** Calls frequently, e.g., *khreeek-khreeek-khreeek* ..., being especially noisy when breeding (*kew-kew-kew-kew*...). **Where to see** Ventanas Bay, Rio Maipo wetland (Valparaíso Region), Chacao Channel (Los Lagos Region), road to Fuerte Bulnes, Puerto Natales Bay (Magallanes Region).

Common Tern
Sterna hirundo   M
L 31–40cm. Sexes identical. **ID** Similar to other *Sterna* terns. Only observed in non-breeding plumage in Chile. Outer primaries usually worn and blackish, contrasting with fresh inner primaries and secondaries (especially in austral spring and early summer). Well-defined blackish carpal patch. Rump pale grey in flight, but beware effects of light. **Habitat** Coasts. **Voice** Silent in Chile, being very vocal only on the breeding grounds. **Where to see** Mainly northern Chile, at Lluta wetland, Arica coast (Arica y Parinacota Region); less regularly in central Chile, at Valparaíso Bay, Rio Maipo wetland (Valparaíso Region).

Antarctic Tern
Sterna vittata   R
L 34–41cm. Sexes identical. **ID** Similar to other *Sterna* terns. Has a rounded body, fine bill and short legs. Rump white in flight, but beware effects of light. **Habitat** Antarctica during breeding season. Rocky islands, stacks and rocky beaches. Remains close to coast. **Voice** Vocal in flight and at colonies, but less so than congenerics. Most frequently heard is a short, sharp, repeated *kip kip, kip kip*.... **Where to see** Antarctic Peninsula and adjacent islands (Magallanes Region), but wanders to southern South America.

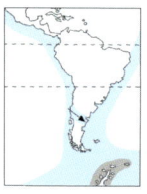

Arctic Tern
Sterna paradisaea   M
L 33–38cm. Sexes identical. **ID** Similar to other *Sterna* terns. In austral spring has heavily worn wing and tail feathers. More rounded head and smaller, more pointed bill than other common *Sterna* terns. Rump white in flight, but beware effects of light. **Habitat** Pelagic, in Antarctic waters (around pack ice) and on migration over open ocean; rarely on mainland coast (usually exhausted birds). **Voice** Up to ten different calls described, but in Antarctica is quiet and infrequently heard; rather short contact calls (*keup – keup – keup*...)**,** or a sharper whistle-like call. **Where to see** Humboldt Current and Antarctic waters (Magallanes Region).

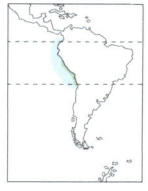

Peruvian Tern
Sternula lorata   R
L 22–24cm. Sexes identical. **ID** Unmistakable. A small tern with long yellow bill, the tip and culmen black. Flight zigzagging, fast and fluttering. Flies up and down same flightpath. **Habitat** Coasts, although probably pelagic at some seasons. **Voice** Some calls very similar to those of other terns, but most frequently heard is a *chirp-chirp-chirp* (recalling a passerine or a sandpiper) and *tcheee-rrriu, tchee-rrriu* (which gives rise to its Chilean name). **Where to see** Antofagasta and Mejillones (Antofagasta Region).

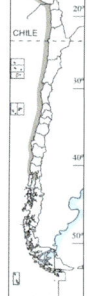

Black Skimmer
Rynchops niger   M
L 40–50cm. Sexes similar. **ID** Unmistakable. Unique bird in Chile, black above and white below, with remarkable bicoloured bill that has longer lower mandible. **Habitat** Coasts, at mouth of rivers. Occasionally a few km offshore. **Voice** Calls frequently, especially flocks in flight, when gives a high-pitched call that resembles a bark (*Kiep ... Kiep ... Kiep*...). Also calls from the ground, a repeated *kaa-kaa-kaakaa*.... **Where to see** Estuaries of the Lluta (Arica y Parinacota Region), Tongoy (Coquimbo Region), Aconcagua, Maipo (Valparaíso Region), and Mataquito (Maule Region). Also Caulín wetland (Los Lagos Region).

PLATE 22: TERNS II

Elegant Tern
Thalasseus elegans **M**
L 39–43cm. Sexes identical. **ID** Unmistakable, perched and in flight. Large tern, with long slightly decurved bill (yellow to orange, intensity varies). Outer primaries worn and blackish, contrasting with fresh inner primaries and secondaries (in austral spring and early summer). Some adults arrive with solid black cap (breeding plumage) but usually have white forehead and black crown and nape, with ragged-looking feathers (non-breeding plumage). Some have orange legs and feet. **Habitat** Coasts, often at estuaries. **Voice** Flocks at rest give a mix of squawks and other calls, but most frequently heard is a short, sharp, persistent cry, given in flight or perched (*kerrreeek ... kerrreeek... kerrreeek... kerrreeek...*). **Where to see** Estuaries, such as Lluta wetland (Arica y Parinacota Region), Rio Maipo wetland (Valparaíso Region), etc. Commoner in central and northern Chile.

Sandwich Tern
Thalasseus sandvicensis **V**
L 36–46cm. Sexes identical. **ID** Unmistakable. A yellow-tipped black bill (sometimes very little yellow). Outer primaries worn and blackish, contrasting with fresh inner primaries and secondaries (in austral spring and early summer). **Habitat** Restricted to coasts in Chile. **Voice** Rather silent during non-breeding season, with the most frequently heard call a short, sharp, persistent cry, given in flight and similar to Elegant Tern (*kirri-kirri-kirri-kirri...*). **Where to see** Estuaries, e.g., Tongoy (Coquimbo Region), Culebrón (Coquimbo Region) and Maipo (Valparaíso Region), with flocks of other terns.

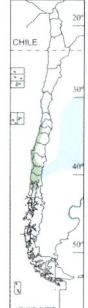

Snowy-crowned Tern
Sterna trudeaui **R**
L 33–36cm. Sexes identical. **ID** Unmistakable. Blackish mask, unique among terns in Chile. Two-toned bill. **Habitat** Coasts, especially estuaries, preferring fresh over salt water. Also lakes, ponds and calm inland rivers. **Voice** Noisy, especially when breeding and in flight. Most commonly heard is a short, very sharp and quickly repeated *keek-keek-keek-keek-keek-keek...* and a raspy, rather low *Kaaeeek… Kaaeeek*. **Where to see** Rio Maipo wetland in winter (Valparaíso Region), Salinas de Cáhuil (O'Higgins Region), Budi lake (Araucanía Region).

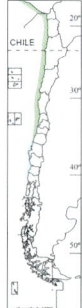

Inca Tern
Larosterna inca **R**
L 39–42cm. Sexes identical. **ID** Unmistakable. Grey body, intensely red bill and striking white 'moustaches'. Juvenile all brown, without 'moustaches', which develop and turn white gradually. **Habitat** Coasts, especially rocky shores and ports. Rarely strays far out to sea. **Voice** Calls frequently, albeit much less at sea than when breeding, including a repeated rather neutral syllable, repeated variably (*kaek-kaek-kaek-kaek-kaek...*), and a more acute aggressive-sounding *Kraeeeck ... Kraeeeck... Kraeeeck....* **Where to see** Arica port (Arica y Parinacota Region), Valparaíso Bay, San Antonio port (Valparaíso Region).

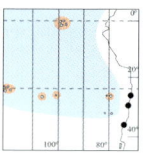

Sooty Tern
Onychoprion fuscatus **R**
L 36–45cm. Sexes identical. **ID** Similar to Grey-backed Tern. Upperparts black, underparts and forehead white. Flight light and buoyant. **Habitat** Pelagic, mostly in tropical Pacific. Also temperate seas near tropics. **Voice** Silent at sea, but very noisy at colonies, giving various high-pitched calls, mostly strident, others drawn-out, scratchy and aggressive-sounding. **Where to see** Easter Island, San Félix and San Ambrosio Islands, and Salas y Gómez Island (Valparaíso Region).

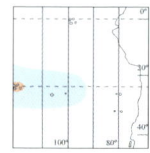

Grey-backed Tern
Onychoprion lunatus **R**
L 34–38cm. Sexes identical. **ID** Similar to Sooty Tern. Upperparts grey, underparts and forehead white, the latter projecting backwards like an eyebrow. **Habitat** Pelagic, mostly in tropical Pacific Ocean. Also temperate waters near tropics. **Voice** Silent at sea, but gives high-pitched, shrill calls at colonies. **Where to see** Easter Island (Valparaíso Region).

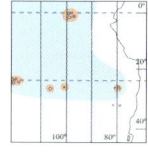

Brown Noddy
Anous stolidus **R**
L 36–45cm. Sexes identical. **ID** Unmistakable. A dark brown tern, with very pale grey forehead and crown. Primaries darker than mantle. **Habitat** Pelagic, in tropical waters. Also temperate waters near tropics. **Voice** Silent at sea, but vocal around colonies; hoarse, raspy calls on nest or in flight (*ghaek-ghaek-ghaek-ghaek-ghaek...*). **Where to see** Easter Island, San Félix and San Ambrosio Islands, and Salas y Gómez Island (Valparaíso Region).

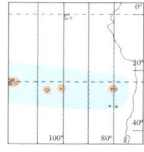

Grey Noddy
Anous albivitta **R**
L 27–30cm. Sexes identical. **ID** Unmistakable. A rather small, predominantly pale grey-plumaged tern, with a rather long-looking, notched tail and pale trailing edge to the wing. **Habitat** Pelagic, but not over deep waters. Always around islands where it breeds. Tropical and warm-water seas. **Voice** Silent at sea, but around nests gives almost melodic chirps, and a kind of purring when perched. If disturbed, calls frequently in flight. **Where to see** Easter Island, and San Félix and San Ambrosio Islands (Valparaíso Region).

White Tern
Gygis alba **R/M?**
L 25–30cm. Sexes identical. **ID** Unmistakable. Large-eyed, all-white tern (not always obvious at a distance) with a relatively chunky black bill, short notched tail and large-headed appearance. **Habitat** Pelagic, mainly in tropical waters. Post-breeding disperses at sea, but remains relatively close to breeding islands. **Voice** Silent at sea, but vocalises frequently on and over land, mainly sharp calls, but some more guttural (*krrik-krrik-krrik-krrik-krrik...*) or (*heek-heek-heek…*). **Where to see** Easter Island and Salas y Gómez Island (Valparaíso Region). **Note** Breeding records at Easter and Salas y Gomez Islands are from the past. The current situation is unclear. [Alt. Common White Tern]

PLATE 23: GULLS IN FLIGHT

PLATE 24: TERNS IN FLIGHT

PLATE 25: SWANS AND WHISTLING DUCKS

Black-necked Swan
Cygnus melancoryphus   R
L 120–122cm. Sexes similar. **C & S** 0–2,600m. **A** 0–1,000m. **ID** Unmistakable. Male has long curved neck; female a straighter, shorter and thicker neck. **Habitat** Inland waters and coasts with abundant vegetation. **Voice** Mainly silent, but gives alarm or contact notes. Best known is a monosyllabic, repeated, sharp *peew-peew-peew-peew-peew*. **Where to see** El Peral lagoon (Valparaíso Region), Cruces River (Los Ríos Region), Chacao and Caulín (Los Lagos Region), Puerto Natales Bay (Magallanes Region) and Torres del Paine National Park (Magallanes Region).

Coscoroba Swan
Coscoroba coscoroba   R
L 108–120cm. Sexes similar. 0–1,500m. **ID** Unmistakable. Male has long curved neck; female a straight, short and thick neck. **Habitat** Inland water with abundant vegetation and nearby grass. Also on coasts. **Voice** Not very vocal. Female gives an onomatopoeic *kooc-ko-roa*. Male gives low-pitched calls. **Where to see** Estero Pachingo (Coquimbo Region), El Yali National Reserve and Cartagena lagoon (Valparaíso Region), Cáhuil Saltworks (O'Higgins Region), Budi lake (Araucanía Region), Puerto Natales Bay and Torres del Paine National Park (Magallanes Region).

White-faced Whistling Duck
Dendrocygna viduata   V
L 45cm. Sexes identical. 0–700m. **ID** Shape like other whistling ducks, but colour diagnostic. Chiloe Wigeon has white foreface, but different shape and body coloration. White face, with contrasting eyes. **Habitat** Wetlands of all types. **Voice** A noisy duck, calling from the ground or in flight, a trisyllabic *see-re-rew*. Female higher-pitched than male. **Where to see** Records have increased during the last decade, from the northern border with Peru to Concepción (Bio-Bio Region).

Fulvous Whistling Duck
Dendrocygna bicolor   V
L 45–49cm. Sexes identical. 0–700m. **ID** Like other whistling ducks, but has face and neck ochre-brown, and a grey bill. **Habitat** Wetlands of all types. **Voice** Noisy, with the most frequently heard call disyllabic, strident and metallic (*pee-tew*), with second syllable sometimes prolonged, given both at rest and in flight. **Where to see** The few records are concentrated in central Chile, with another at Lluta wetland (Arica y Parinacota Region).

Black-bellied Whistling Duck
Dendrocygna autumnalis   V
L 47–51cm. Sexes identical. 0–700m. **ID** Like other whistling ducks, but has grey face and pinkish to reddish bill. In flight the only whistling duck with a large white area on the upperwing-coverts. **Habitat** Wetlands with surrounding trees. **Voice** Calls often, especially in flocks. The most typical call is given on the ground or in flight, a shrill whistle of four syllables, with the first more intense (*twi-twi-ri-ri-ri*). **Where to see** Recorded from northern border with Peru to Osorno (Los Lagos Region).

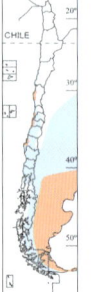
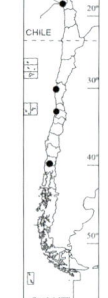
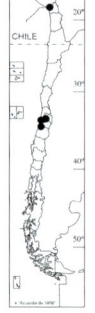
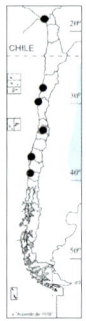

PLATE 26: GEESE

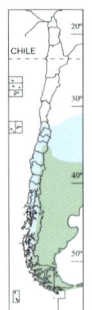

Upland Goose
Chloephaga picta R
L 66–75cm. Sexually dimorphic. **C** 2,500–3,500m. **S & A** 0–2,000m. **ID** Male unmistakable. Female similar to Ruddy-headed Goose, but relatively larger, with brownish and very prominent forehead. Barred body with thick stripes. Both sexes have two morphs. **Habitat** In central Chile, Andean wetlands; in the south, varied environments, except forest, including coasts, wet meadows, Patagonian steppe and inland waters. **Voice** Male high-pitched and female low-pitched. Most frequently heard is territorial call of male, a sustained, high-pitched whistle (*few, few, few few, few, few, few...*) and a shrill whistle (*weeou*). Female usually gives a nasal cackle and quiet *kha, kha, kha...*. **Where to see** Yeso Valley (Metropolitan Region), Torres del Paine National Park and Pali Ayke National Park, and road to Fuerte Bulnes (Magallanes Region).

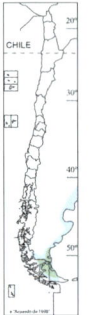

Ruddy-headed Goose
Chloephaga rubidiceps R
L 55–60cm. Sexes identical. **S & A** 0–1,000m. **ID** Similar to female Upland Goose, but smaller, with greyish body and without prominent forehead. Barred body, with fine stripes in both sexes. **Habitat** Diverse environments, except forest: coasts, wet meadows, Patagonian steppe and inland waters. **Voice** Male high-pitched and female gives rough, nasal and low-pitched notes. Mostly heard in alarm, a loud sharp *pew-pew-pew-pew-pew...* (from the ground or in flight) and a lower *kha, kha, kha, kha...*. **Where to see** Forms mixed flocks with other geese. Road to Fuerte Bulnes, Buque Quemado and Pali Ayke National Park (Magallanes Region).

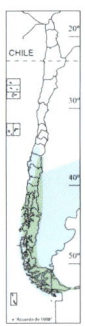

Ashy-headed Goose
Chloephaga poliocephala R
L 60–62cm. Sexes identical. **S & A** 0–2,000m. **ID** Adult unmistakable. Juvenile has greyish head and whitish belly. **Habitat** Steppe near sea, bodies of water, rivers and estuaries, preferably close to forests. **Voice** Does not vocalise often, except when breeding and in alarm. Male gives sibilant, high-pitched *fee-fee-fee...*, probably thinner than other geese, and female utters a nasal or rough-sounding *hak-hak-hak...*. **Where to see** Conguillio National Park (Araucanía Region), road to Fuerte Bulnes and Torres del Paine National Park (Magallanes Region).

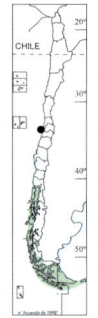

Kelp Goose
Chloephaga hybrida R
L 60–72cm. Sexually dimorphic. **S & A** 0–50m. **ID** Unmistakable. Always seen in a couple. Adult male all white (very exposed), with small, dark bill. Adult female mainly blackish with barred underparts (usually hidden), pale eye-ring and pinkish bill. Both sexes have yellow legs. Young male is mainly female-like with white head. **Habitat** Coasts, rocky or stony, with small algae. **Voice** Male high-pitched and sibilant, and female low-pitched, harsh and grunting, but both sexes can sound higher-pitched in alarm. Another alarm call is similar, but very low and is apparently given only by female (*hak-hak-hak-hak-hak-hak...*), as well as high-pitched whistles (*fee-fee-fee-fee...*), apparently only by male. Chicks give short high-pitched whistles, similar to others of genus. **Where to see** Puñihuil (Los Lagos Region), road to Fuerte Bulnes and Chilota Bay, Porvenir (Magallanes Region).

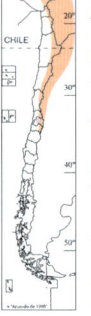

Andean Goose
Chloephaga melanopterus R
L 77cm. Sexes similar. **N** 3,500–5,000m. **C** 600–4,000m. **ID** Unmistakable. Unique black-and-white pattern seen only in this distribution range. Dark-tipped pink bill and bright reddish-pink legs. **Habitat** In northern Chile, bogs, salt flats and waterbodies; in central Chile, Andean wetlands and humid valleys. **Voice** Male high-pitched and sibilant, female lower, harsher and somewhat grunting. Regularly heard is short, frequently repeated *kreeep-kreeep-kreeep-kreeep-* ... in alarm, while walking slowly away or in flight. Also a metallic, repeated *kak-kak-kak-kak-kak-kak...*, apparently with a territorial function. **Where to see** Parinacota (Arica y Parinacota Region), Machuca (Antofagasta Region), El Yeso Valley and Lampa/Batuco system, mostly in winter (Metropolitan Region).

PLATE 27: DUCKS I

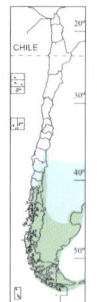

Flying Steamerduck
Tachyeres patachonicus R

L 65–70cm. Sexually dimorphic. **S & A** 0–1,500m. **ID** Similar to Magellanic Steamerduck, but smaller. Wings almost cover the back and tail rather long. Female has purple/maroon-grey head and orange-grey bill. **Habitat** Wetlands, both coastal and inland. The only steamerduck found at inland lakes. **Voice** Calls often, especially in territory defence, or intra/interspecific aggression. Male has wide repertoire, including high-pitched whistles (*krreew-krreew-krreew…*), various grunts including harsh sounds like metal scraping a stone, and other low-pitched noises. Female is lower-pitched, also uttering much more audible growls, which are repeated several times. **Where to see** Llanquihue lake (Los Lagos Region), Torres del Paine National Park (Magallanes Region).

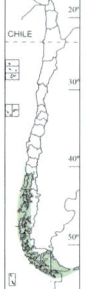

Magellanic Steamerduck
Tachyeres pteneres R

L 78–84cm. Sexually dimorphic. **ID** Similar to Flying Steamerduck, but larger. Short wings do not cover back, and tail is relatively short. Two morphs (southern and austral), which differ in bill colour, especially females. **Habitat** Coasts and fjords. Never on inland waters not connected to the sea. **Voice** Calls frequently. Male usually high-pitched, and the female typically giving lower, usually rough, nasal grunts. **Where to see** Caulín and Puñihuil (Los Lagos Region), road to Fuerte Bulnes and Porvenir (Magallanes Region). [Alt. Fuegian Steamerduck, Flightless Steamerduck]

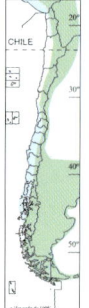

Crested Duck
Lophonetta specularioides R

L 50–60cm. Sexes similar. **N** 2,500–4,600m. **C** 2,500–3,.500m. **S & A** <1,000m. **ID** Unmistakable. Note head shape. Orange eyes in northern Chile, intense red in south and austral. **Habitat** High-Andean wetlands in the north and central Chile. Coasts and inland waters in austral zone. **Voice** Pairs defend territory aggressively, with male giving various high-pitched notes, and female 'severe', nasal and rough sounds. In response threat, pair reacts collaboratively, simultaneously giving a shrill *pew pew pew pew* (male) and a low, raspy *quaek-quaek-quaek…* or *kek-kek-kek-kek-kek…* (female). **Where to see** Lauca National Park (Arica y Parinacota Region), Atacama Salt Flat (Antofagasta Region), El Yeso Reservoir (Metropolitan Region), Tres Puentes wetland and Puerto Natales Bay (Magallanes Region).

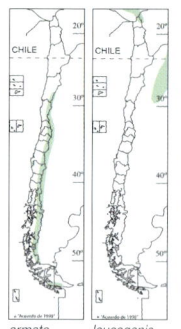

Torrent Duck
Merganetta armata R

L 40–46cm. Sexually dimorphic. **N** 3,700–4,300m. **C & S** 800–3,000m. **A** 50–1,000m. **ID** Unmistakable. Beautiful red-billed duck; adult male with boldly black-and-white-patterned head, and otherwise mainly dark plumage, and adult female mainly grey above with chestnut underparts. **Habitat** Rivers and streams with clear or turbid water, even sometimes with little water. **Voice** Rarely heard, or inaudible due to the background noise of water. Male high-pitched, like a hiss and, sometimes,

armata   *leucogenis*

a whistle. Female lower and raspier, similar to other ducks (*queeck-queeck-queeck-queeck…*). **Where to see** Juncal River (Valparaíso Region), road to Termas del Flaco (O'Higgins Region), Maule River (Maule Region), Petrohué River (Los Lagos Region), Pingo River (Magallanes Region).

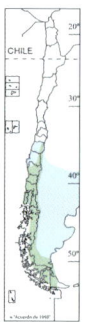

Spectacled Duck
Speculanas specularis R

L 40–53cm. Sexes identical. **S** 800–1,800m. **A** 0–1,000m. **ID** Unmistakable. White facial spot varies in size. **Habitat** Mountain rivers with forested margins. In the south occupies lakes and lagoons, even small, seasonal ones. **Voice** Not frequently heard. Male gives harsh sharp sounds, and a high-pitched *pew-pew-pew….* Female gives a bark (*haek-haek-haek…*) among other calls. **Where to see** La Laja lagoon (Bio-Bio Region), Torres del Paine National Park and Serrano River (Magallanes Region).

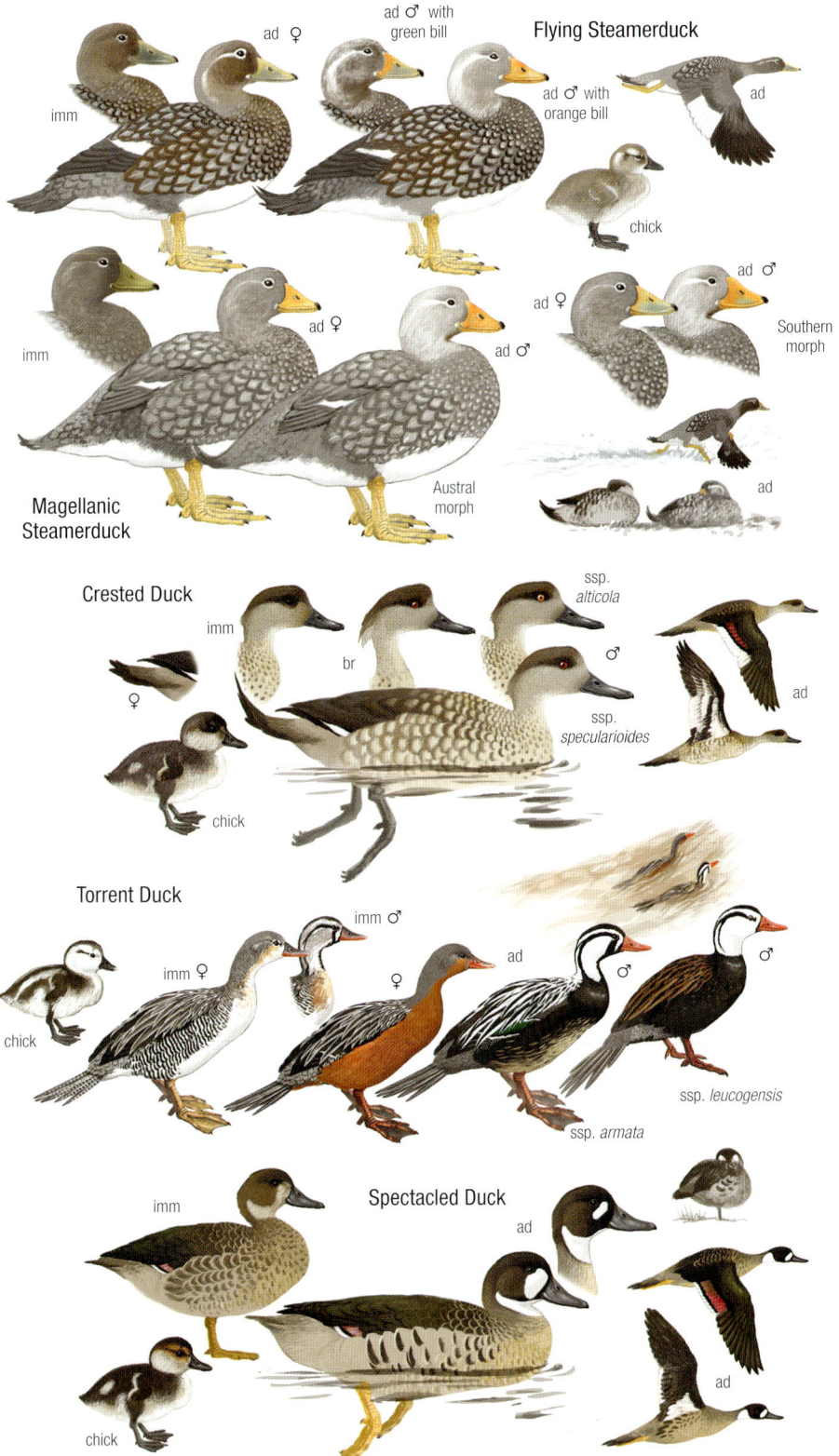

PLATE 28: DUCKS II

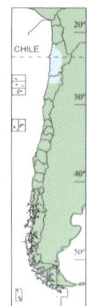

Yellow-billed Pintail
Anas georgica   R
L 48–51cm. Sexes similar. **N** 0–4,600m. **C** 0–3,000m. **S & A** 0–2,000m. **ID** Rounded head shape, similar in colour to rest of body. Eye appears like a black circle. **Habitat** Various wetlands, permanent or temporary, dams and rice fields. Also marine bays. **Voice** Does not vocalise frequently, usually in alarm, a shrill trilled whistle (*prrreeep*), given only by male, and a low-pitched squawk (*queck-queck-queck-queck*...), typical of female. Both calls sometimes given in threat. **Where to see** Any Chilean wetland, from Chungará lake (Arica y Parinacota Region) to Beagle Channel and south (Magallanes Region).

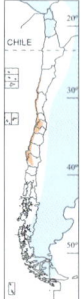

White-cheeked Pintail
Anas bahamensis   R
L 49cm. Sexes similar. **N** 0–4,600m. **C & S** 0–2,000m. **ID** Unmistakable. Cheeks and upper neck white. **Habitat** Wetlands of all types. **Voice** Largely silent, but male has a multisyllabic sharp call that resembles a child's whistle, and female gives quieter and low-pitched calls. **Where to see** Lluta wetland (Arica y Parinacota Region), Lampa/Batuco system (Metropolitan Region).

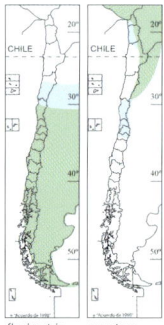

Yellow-billed Teal
Anas flavirostris   R
L 38–43cm. Sexes identical. **N** 800–4,600m. **C** 0–3,000m. **A** 0–1,000m. **ID** Unmistakable. Head appears square and dark, marked by a dark eyestripe. **Habitat** Any wetland, permanent or temporary. Also coastal bays (in south). **Voice** Generally rather silent. Mostly heard is a trilled whistle in alarm (*prrreeep*), sharp and not very powerful, apparently only by male, and a low-pitched *queck* by female, sometimes given simultaneously when faced by a threat. **Where to see** Lauca National Park (Arica y Parinacota Region), El Peral lagoon (Metropolitan Region), Lampa/Batuco system (Metropolitan Region), Chiloé Island (Los Lagos Region), Puerto Natales Bay (Magallanes Region).

flavirostris   *oxyptera*

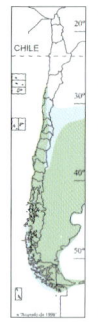

Chiloe Wigeon
Mareca sibilatrix   R
L 51cm. Sexes similar. **C, S & A** 0–1,000m. **ID** Unmistakable. White-faced Whistling Duck has white foreface, but body has different shape and coloration. Eyes situated within dark feathers (unlike the whistling duck), and iridescent head. **Habitat** Various wetlands, permanent or temporary. Also coastal bays (in south). **Voice** Very vocal. Male gives characteristic whistle of two syllables (*fuuuiiit-tewwwww*) likened to a human whistle, given constantly in alarm or flight. Female has a rougher, nasal voice. Flocks on water call constantly. **Where to see** El Peral lagoon (Metropolitan Region), Lampa/Batuco system (Metropolitan Region), Llanquihue lake and Chiloé Island (Los Lagos Region), Puerto Natales Bay (Magallanes Region).

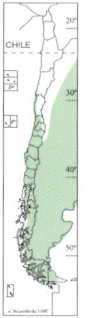

Red Shoveler
Spatula platalea   R
L 51cm. Sexually dimorphic. **C & S** 0–800m. **ID** Male unmistakable. Distant female recalls other brown ducks, but has a dark obviously large bill. **Habitat** Lowland lakes, lagoons or swamps, with submerged vegetation. Never at Andean lagoons. **Voice** Soft sounds, rarely audible except in flight, alarm or aggression. **Where to see** El Peral lagoon (Valparaíso Region), Lampa/Batuco system (Metropolitan Region), Rapel lake (O'Higgins Region), Budi lake (Los Lagos Region), Torres del Paine National Park (Magallanes Region).

PLATE 29: DUCKS III

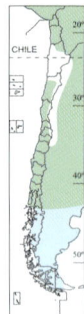

Cinnamon Teal
Spatula cyanoptera R
L 43–48cm. Sexually dimorphic. **N** 0–4,600m. **C** 0–800m. **ID** Male unmistakable. At long range, female similar to other dark-billed female ducks, such as Red Shoveler, but bill base usually pale, or spotted pale; bill is proportionate to the body. **Habitat** Lagoons, lakes or rivers. **Voice** Mainly calls in alarm. Male gives quiet, high-pitched, sibilant *kho, kho, kho* ... or is very similar to other dabbling ducks, a musical *queck, queck, queck*. Female less high-pitched, hoarser and more grunting (*gaack, gaack, gaack*...). **Where to see** Lluta wetland (Arica y Parinacota Region), Loa River, from Calama to coast (Antofagasta Region), Lampa/Batuco system (Metropolitan Region).

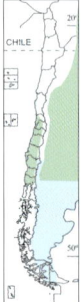

Rosy-billed Pochard
Netta peposaca R
L 50–55cm. Sexually dimorphic. **C, S & A** 0–800m. Uncommon. **ID** Unmistakable male. Female recalls other brown ducks with dark bills, but white undertail-coverts diagnostic, even at long range. **Habitat** Freshwater wetlands. Not in Andes. **Voice** Usually silent, except in alarm or flight, when gives a hoarse croak, drawn-out like a moan (*graaaack, graaaack, graaaack*) and a long growl (*grrrrrrck-grrrrrrck*), perhaps by both sexes, but probably the female alone. Also sharp, short calls, like whistles. **Where to see** Lampa/Batuco system (Metropolitan Region), road Punta Arenas to Puerto Natales lagoon (Magallanes Region).

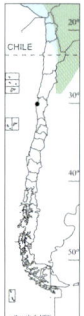

Puna Teal
Spatula puna R
L 45–49cm. Sexes similar. **N** 0–4,600m. **ID** Unmistakable. Very similar to Silver Teal, but no overlap. White cheeks and pale blue bill. **Habitat** Wetlands of all types, including rivers, mainly in high Andes, but not unusual at sea level. **Voice** Generally quiet whistles, squawks and grunts, always soft. Most frequently heard in flight is a drawn-out, somewhat agonised *queeeeck-queeeeck-queeeeck*.... In alarm, female gives a four-part call, halfway between a croak and growl (*quack-quack-quack-quack*). **Where to see** Parinacota/Cotacotani system and Chungará lake (Arica y Parinacota Region), Putana and Loa River, approximately from junction with San Salvador River to sea (Antofagasta Region).

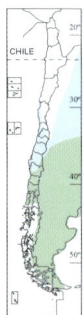

Silver Teal
Spatula versicolor R
L 40–43cm. Sexes similar. **C, S & A** 0–1,000m. **ID** Unmistakable. Very similar to Puna Teal, but no overlap. Ochre cheeks and yellow spot at base and on upper bill. **Habitat** Freshwater, shallow wetlands. Very local. **Voice** Calls low and somewhat nasal, only becoming louder in alarm. Most frequently heard is a cricket-like *krrrek, krrrek, krrrek* ..., or typical duck *queck, queck, queck, queck*... and a drawn-out, slightly agonised *kwooooc, kwooooc, kwooooc* ..., given repeatedly in flight. **Where to see** Buque Quemado wetland (Magallanes Region).

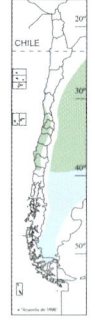

Black-headed Duck
Heteronetta atricapilla R
L 40–42cm. Sexually dimorphic. **C & S** 0–600m. **ID** Male unmistakable. Bluish bill with red spot at base. Distant female resembles other dark-billed brown ducks. Has pale eyestripe and compact appearance, with very short tail and neck usually retracted. **Habitat** Freshwater wetlands in lowlands. **Voice** Usually calls only from within cover, and female hardly at all. Most calls of male are low grunts, including an accelerating, husky and low-pitched *kho kho kho kho*. Also a high-pitched alarm, like a whistle (*wheet*). **Where to see** Lampa/Batuco system (Metropolitan Region), El Peral lagoon and Campiche-Ventanas wetland (Valparaíso Region).

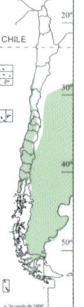

Lake Duck
Oxyura vitatta R
L 37–39cm. Sexually dimorphic. **C & S** 0–1,000m. **ID** Similar to Andean Duck. Male has neck completely black. Female has thick whitish lines on cheek. **Habitat** Freshwater wetlands in lowlands. **Voice** Silent except when breeding. **Where to see** El Peral lagoon (Valparaíso Region), Lampa/Batuco system (Metropolitan Region), lakes near Villarica (Araucanía Region), Budi lake (Araucanía Region).

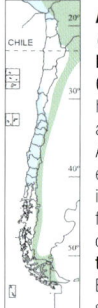

Andean Duck
Oxyura ferruginea R
L 41–47cm. Sexually dimorphic. **N** 2,000–4,600m. **C, S & A** 0–2,500m. **ID** Similar to Lake Duck. Male has foreneck chestnut, hindneck black. Female has all-dark cheeks. **Habitat** Freshwater wetlands. High-Andean wetlands in the north. **Voice** Generally silent, especially female. Male displays to female by raising its body and paddling in water, while jabbing towards female with its bill, finally giving a single monosyllabic call like a toad, before the display is repeated. **Where to see** Chungará lake (Arica y Parinacota Region), El Peral lagoon (Valparaíso Region), Lampa/Batuco system (Metropolitan Region), lakes near Villarica (Araucanía Region), Torres del Paine National Park (Magallanes Region). **Note** Formerly treated as conspecific with Ruddy Duck *O. jamaicensis*.

PLATE 30: GREBES

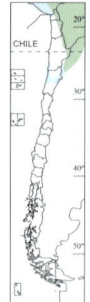

Northern Silvery Grebe
Podiceps juninensis R
L 28cm. Sexes identical. **N** 0–4,600m. **ID** Only confusable with Southern Silvery Grebe, although the two are rarely seen in the same places. The white throat is the main difference from Southern. **Habitat** High-Andean waterbodies, occasionally on the sea. Breeds also on the Loa River. **Voice** A varied repertoire of short and sharp calls, usually only heard when breeding. **Where to see** Chungará and Parinacota lakes (Arica y Parinacota Region), Loa River (Antofagasta Region). **Note** Formerly considered to be a subspecies of Southern Silvery Grebe, using the name Silvery Grebe for the combined species.

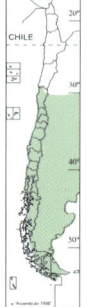

Southern Silvery Grebe
Podiceps occipitalis R
L 28cm. Sexes identical. **C, S & A** 0–2,600m. **ID** See Northern Silvery Grebe, but the present species has a grey (not white) throat. At long range, and only in far south, might also be confused with the very rare Hooded Grebe. Grey throat and forehead useful in all plumages. **Habitat** Interior waterbodies, even in Andes. Also frequently on sea, especially in winter. **Voice** A varied repertoire of short high-pitched calls are mainly heard only when breeding; otherwise typically silent. **Where to see** El Peral lagoon (Valparaíso Region), Batuco/Lampa system (Metropolitan Region), Rio Maipo wetland (Valparaíso Region), Llanquihue lake (Los Lagos Region), Torres del Paine National Park (Magallanes Region).

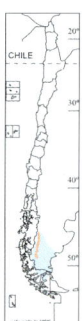

Hooded Grebe
Podiceps gallardoi V
L 33–35cm. Sexes identical. 0–1,000m. **ID** At long range could be confused with Southern Silvery Grebe. Rufous in crown and black throat. **Habitat** Very local in Patagonia; freshwater lagoons with abundant floating vegetation, especially 'vinagrilla' (*Myriophyllum quitense*). **Voice** No information. **Where to see** No regular localities in Chile.

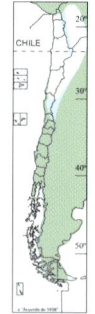

White-tufted Grebe
Rollandia rolland R
L 20–26cm. Sexes identical. **N** 0–4,600m. **C, S & A** 0–1,200m. **ID** Unmistakable. Breeding adult has dark head, neck and upperparts, rufous flanks and striking fan-shaped white ear plumes, which are less obvious in non-breeding plumage, at which season neck and underparts are duller rufous. **Habitat** Fresh or saltwater wetlands; marine bays and fjords. **Voice** Several calls, of which a low, monosyllabic *kek*, given repeatedly, and a high-pitched, monosyllabic *kwiw*, also repeated several times, are perhaps most frequently heard. **Where to see** Chungará lake (Arica y Parinacota Region), Batuco/Lampa system (Metropolitan Region), El Peral and Cartagena lagoons (Valparaíso Region), Torres del Paine National Park (Magallanes Region).

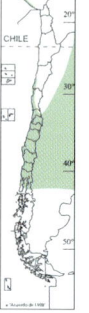

Pied-billed Grebe
Podilymbus podiceps R
L 30–31cm. Sexes identical. 0–800m. **ID** Unmistakable. Elongated head and relatively large bill crossed by black stripe. **Habitat** Freshwater bodies with fish and/or amphibians and invertebrates. Occasionally artificial lagoons in urban areas. **Voice** Usually calls only from within cover, so rarely seen vocalising. Best known is a long repetitive sequence, which diminishes towards end (*kó-kó-kó-kó-kó ... kó ... .kó ... .kó ... ..kó*) and has a metallic timbre. **Where to see** Catapilco River, Cartagena lagoon, El Peral lagoon (Valparaíso Region).

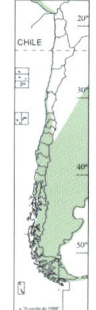

Great Grebe
Podiceps major R
L 60–78cm. Sexes identical. 0–1,200m. **ID** Unmistakable. Silhouette unique in ponds and lakes. Very long-necked, and long, sharply-pointed billed grebe. Head largely black in breeding plumage, but mostly whitish-grey in non-breeding season, when neck is paler and duller rufous, and bill also much paler. **Habitat** Freshwater bodies. Also marine bays and austral fjords, sometimes even a few km offshore. **Voice** Largely quiet, but gives a monosyllabic call, like a plaintive moan, at twilight, commonly from the middle of a wetland (similar to the call of a diver/loon), as well as a high-pitched repeated *kiip-kiip-kiip-kiip-kiip-kiip* ..., apparently in alarm. **Where to see** Mantagua wetland (Valparaíso Region), Lampa/Batuco system (Metropolitan Region), Aculeo lagoon (O'Higgins Region), Llanquihue lake (Los Lagos Region), Torres del Paine National Park, ponds near Porvenir (Magallanes Region).

PLATE 31: COOTS

All coots share a blackish body, making it essential to check the colour of the bill and frontal shield to correctly identify the species. Present in almost any wetland in range, permanent or seasonal.

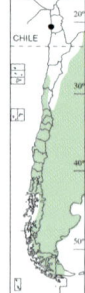

Red-gartered Coot
Fulica armillata **R**
L 50–55cm. Sexes identical. **C & S** 0–2,600m. **ID** Bill and frontal shield are both yellow. Frontal shield is pointed, and separated by a red band at base of bill. Undertail-coverts show little white. **Habitat** Any freshwater body, with or without fringing vegetation. **Voice** A short whistle-like *wheet* is mostly heard, along with a hoarse cackling *ko-ko-ko-ko-ko*. **Where to see** El Peral lagoon (Valparaíso Region), Lampa/Batuco system, Cartagena lagoon (Valparaíso Region).

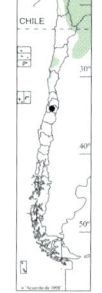

Andean Coot
Fulica ardesiaca **R**
L 46–50cm. Sexes identical. 0–4,600m. **ID** Similar to White-winged Coot. Frontal shield can be white, yellow, or burgundy (maroon), with the latter commonest in Chile. Those with white frontal shield have grey legs, those with burgundy shield have green legs. Note bill shape. **Habitat** Lakes and Andean lagoons with abundant submerged vegetation. Also calm backwaters of estuaries and rivers, exceptionally coastal waters. **Voice** Similar to other coots, but usually low-pitched and quiet. Most usual is short, sharp monosyllabic alarm, like metal being struck (*keek*). **Where to see** Chungará lake (Arica y Parinacota Region), Putana River, Loa River and Sloman dam (Antofagasta Region). [Alt. Slate-coloured Coot]

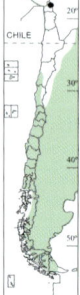

White-winged Coot
Fulica leucoptera **R**
L 42–44cm. Sexes identical. 0–2,600m **ID** Similar to Andean Coot, but has pale yellow bill, with frontal shield same colour or, in some cases, orange. Also note bill shape. Undertail-coverts partially white. **Habitat** Any freshwater body, preferably with fringing vegetation. Also coastal bays in south of range. **Voice** Several different calls, all of which are similar to those of other sympatric coots, although its high-pitched monosyllabic alarm call (*quiek*) is reasonably distinctive. **Where to see** El Peral and Cartagena lagoons (Valparaíso Region), Lampa/Batuco system (Metropolitan Region).

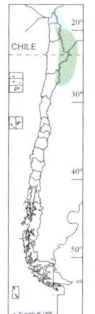

Horned Coot
Fulica cornuta **R**
L 51–62cm. Sexes identical. 3,000–4,500m. **ID** Unmistakable. Unique frontal appendage that resembles a 'horn', and which replaces frontal shield in both sexes. **Habitat** Lakes and lagoons with abundant submerged vegetation; also in backwaters of rivers, all in high Andes. **Voice** Similar to other coots, generally a short *queek* or *hoot-hoot* like grunting or snorting. **Where to see** Miscanti and Miñique lagoons, Los Flamencos National Reserve (Antofagasta Region), Santa Rosa lagoon (Atacama Region).

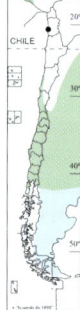

Red-fronted Coot
Fulica rufifrons **R**
L 46–48cm. Sexes identical. 0–2,600m **ID** Yellow bill and dark red frontal shield, the latter long and pointed on crown. Undertail-coverts extensively white. **Habitat** Any freshwater body, preferably with fringing vegetation. **Voice** Within cover utters a hoarse cackling *kóoo-ko-ko-ko....* In alarm, a high-pitched whistle. **Where to see** El Peral lagoon (Valparaíso Region), Lampa/Batuco system (Metropolitan Region).

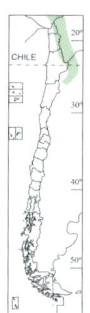

Giant Coot
Fulica gigantea **R**
L 62–66cm. Sexes identical. 3,500–4,600m. **ID** Unmistakable. The largest coot, with dark red bill, concave forehead, and white frontal plate, edged yellow. Adults have two knobs at the top sides of the frontal plate. **Habitat** Highland lakes and lagoons with abundant submerged vegetation; also in backwaters of estuaries and rivers. **Voice** A characteristic hoarse cackling (like other coots) or prolonged whistles, the latter apparently only given by female. Male gives a low *Hurrrrrrrr ... Hurrrrrrrr....* **Where to see** Parinacota wetland and Chungará lake (Arica y Parinacota Region), Putana River (Antofagasta Region).

PLATE 32: GALLINULES AND RAILS

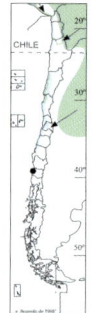

Common Gallinule
Gallinula galeata   R
L 34–37cm. Sexes identical. **N** 0–4,300m. **C** 0–100m. **ID** Unmistakable. Red frontal shield and extensively white undertail-coverts. **Habitat** Wetlands with tall vegetation. Most often at high-Andean bogs (bofedales), coastal lagoons and rivers. **Voice** Several vocalisations, many of them very similar to those of coots. **Where to see** Lluta wetland (Arica y Parinacota Region), Isluga (Tarapacá Region), Highway 5 at where it crosses Loa River (Antofagasta Region), Humedal El Culebrón (Coquimbo Region).

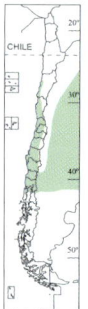

Spot-flanked Gallinule
Gallinula melanops   R
L 28–30cm. Sexes identical. **C & S** 0–800m. **ID** Unmistakable. Yellowish-green bill and shield, and white freckles on flanks. **Habitat** Freshwater bodies with surrounding vegetation. Even urban parks. **Voice** Mostly silent and mainly heard at night, albeit not exclusively. Most common call is like a metallic laugh, but other calls like those of coots. Usually heard while concealed in reeds, rather than in the open waters, although pairs occasionally call in the open to proclaim their territory. **Where to see** El Peral and Cartagena lagoons (Valparaíso Region), Leyda lagoons (Metropolitan Region), Bicentenario Park (Metropolitan Region).

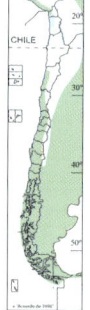

Plumbeous Rail
Pardirallus sanguinolentus   R
L 38–40cm. Sexes identical. **N** 0–4,000m. **C, S & A** 0–800m. **ID** Unmistakable. Yellowish-green bill with red spot at base, red legs and dark undertail-coverts. Juvenile has brownish body and blackish bill and legs. **Habitat** All types of wetlands, e.g. lagoons, lakes, rivers, estuaries and irrigation channels, provided there is fringe vegetation. **Voice** Regularly heard during breeding season, when several individuals may call together, in a chorus lasting several minutes; a characteristic *reeeewr-wheet*. **Where to see** El Peral lagoon (Valparaíso Region), Lampa/Batuco system (Metropolitan Region), Torres del Paine National Park (Magallanes Region), almost any wetland in range.

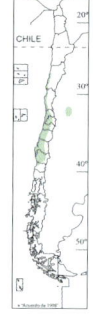

Black Rail
Laterallus jamaicensis   R
L 14–16cm. Sexes identical. **C & S** 0–600m. **ID** Unmistakable if seen well, but otherwise can be confused with the young of other rails, or even a rodent, when seen quickly running between cover. **Habitat** Humid areas with dense vegetation. Muddy or flooded areas, coastal salt pans and even irrigation ditches. **Voice** Quiet, being regularly heard only during breeding season. Two main vocalisations: the territorial song, given day and night (*ki-kee-kee-kee-dúúú*), and a dull-sounding, low-pitched growl (*grrr-grrr-grrr...*). **Where to see** More often heard than seen. Elqui River drainage (Coquimbo Region), Lampa/Batuco system (Metropolitan Region).

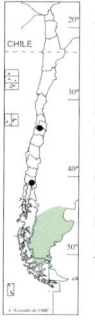

Austral Rail
Rallus antarcticus   R
L 22cm. Sexes identical. 0–500m. **ID** Unmistakable. Red bill and white undertail-coverts. Brown mantle, streaked black. **Habitat** Lakes with reeds (*Juncus* sp., *Scirpus* sp., *Typha* sp.) in Patagonia. **Voice** Generally silent, but during breeding season calls at nightfall. One is a low call, decreasing in intensity (*ua-ua-ua-ua ...*), but others are very sharp, recalling metal beating on metal (*pi-tik…pi-tik…. pi-tik…. pi-tik…*). **Where to see** Very local in southern Chile, Torres del Paine National Park (Magallanes Region).

Purple Gallinule
Porphyrio martinicus   V
L 30cm. Sexes identical. **N** 0–800m. **ID** Unmistakable adult is mainly greenish above with purplish-blue neck and underparts, plus red-and-yellow bill, and yellow legs. Juvenile (more likely to be seen in Chile) is mainly brownish, with darker upperparts and dull blue-green wings, and shares yellow legs and white vent with adult. **Habitat** Flooded areas with abundant vegetation. **Voice** Very vocal, often with a metallic timbre. Probably most often heard is a short monosyllabic *keek*, sometimes repeated (*Keek ... Keek ... keek ... keek ... keek*). Other sounds recall those of coots. **Where to see** Recorded at the Lluta wetland (Arica y Parinacota Region).

PLATE 33: FLAMINGOS AND SPOONBILL

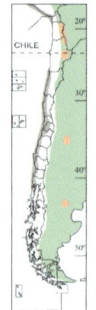

Chilean Flamingo
Phoenicopterus chilensis   R
L 106–125cm. Sexes identical. **N** 0–4,600m. **C** 0–2,200m. **S & A** <600m. **ID** Similar to other flamingos. Adult leg and bill colour diagnostic. Due to the white iris, even at distance or in flight, no 'black point' is visible on the head. Juvenile and immature: check curvature of upper mandible and the division between upper and lower mandibles. The only flamingo in Chile with a posterior toe, albeit vestigial. **Habitat** Varied wetlands. In the north, salt and high-Andean lagoons; in central Chile, Andean lakes, coastal lagoons, intertidal marshes or sheltered and shallow coastal bays; in the south, lagoons and marine bays. **Voice** Most vocal when in groups, but calls rather quiet. Mostly short, high-pitched squawks, sometimes grunts. The local onomatopoeic name is based on one call (*toc-ko-kò*). **Where to see** Chungará lake, Surire Salt Flat (Arica y Parinacota Region), Huasco Salt Flat (Tarapacá Region), Atacama Salt Flat (Antofagasta Region), Yali River (Valparaíso Region), Mataquito River in winter (Maule Region), Lenga wetland (Bio-Bio Region), Chiloé in winter (Los Lagos Region), Torres del Paine National Park and inland lagoons near Porvenir (Magallanes Region).

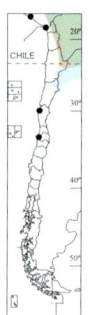

Andean Flamingo
Phoenicoparrus andinus   R
L 100–140cm. Sexes identical. 2,500–4,600m. **ID** Similar to other flamingos. Adult leg and bill colour diagnostic. Some have legs yellowish-tinged. Iris dark, thus even at distance or in flight the eye shows as a 'black point'. Remiges form large black triangle at the end of the body, over the tail. Juvenile and immature: check curvature of upper mandible and the division between upper and lower mandibles. **Habitat** Highland lagoons and lakes. Very rarely on coast. **Voice** Calls frequently in groups, albeit quietly, with growling calls, or a monosyllabic, hoarse, low *kho ... kho ...* in contact, as well as combinations of this, e.g., *kho- kho-kho-kho....* **Where to see** Chungará lake, Surire Salt Flat (Arica y Parinacota Region), Salar de Huasco (Tarapacá Region), Soncor (Chaxa) lagoon (Antofagasta Region), Negro Francisco lagoon (Atacama Region).

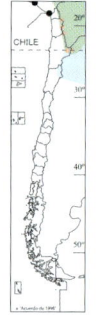

Puna Flamingo
Phoenicoparrus jamesi   R
L 90–110cm. Sexes identical. 2,500–4,600m. **ID** Similar to other flamingos. Adult leg and bill colour diagnostic. Bill shorter and rounder than other species, yellow with only the tip black. Iris dark, thus even at distance or in flight the eye shows as a 'black point'. Juvenile and immature: check bill shape, which differs from both other species. **Habitat** Highland lagoons and lakes. Almost strictly Andean flamingo, with a few records (juveniles) at Lluta wetland. **Voice** Very noisy in flocks, sounding obviously higher-pitched than the other two flamingo species. One call gives rise to its local name (*tchou-ru-ru*). **Where to see** Surire Salt Flat (Arica y Parinacota Region), Huasco Salt Flat (Tarapacá Region), Soncor (Chaxa) lagoon (Antofagasta Region). [Alt. James's Flamingo]

Roseate Spoonbill
Platalea ajaja   V
L 78–81cm. **W** 130cm. Sexes identical. **N & C** <500m. **ID** Unmistakable. At great distance, and initial glance, might be confused with a flamingo, but bill diagnostic. **Habitat** Shallow wetlands. **Voice** Rarely heard guttural voice, slightly strident, even in alarm. **Where to see** Most probably at Lluta wetland (Arica y Parinacota Region).

PLATE 34: HERONS, EGRETS AND NIGHT HERONS

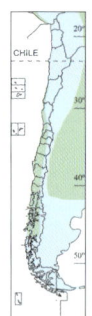

Cocoi Heron
Ardea cocoi   R
L 115–120cm. Sexes identical. **C, S & A** <800m.
ID Unmistakable. The largest Chilean heron, with grey upperparts, black upper head and a yellowish bill (duller in juvenile). **Habitat** Wetlands. In the south, in bays or marine channels, and will perch on tall trees. **Voice** Noisy when breeding, but otherwise rather silent. In flight a hoarse, guttural sound, repeated about four times at decreasing volume (*Haok … Haok … Haok … Haok…*). **Where to see** Lampa/Batuco system (Metropolitan Region), Mantagua wetland (Valparaíso Region), Rapel lagoon (O'Higgins Region), southern fjords (Aysén Region).

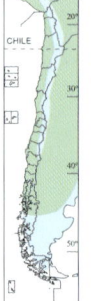

Great Egret
Ardea alba   R
L 85–90cm. **W** 140cm. Sexes identical. **N** 0–4,000m. **C, S & A** <800m. **ID** Unmistakable. Yellow bill with black legs and feet. **Habitat** All types of wetlands. Also arid land far from water. **Voice** A grunt in escape flight (*gaah … gaah*). At rest, and especially when breeding, gives strong *kraak…kraak*. Territorial disputes involve much calling. **Where to see** Lluta wetland (Arica y Parinacota Region), Elqui River (Coquimbo Region), Lampa/Batuco system (Metropolitan Region), Rapel lagoon (O'Higgins Region), etc.

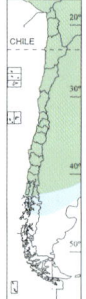

Snowy Egret
Egretta thula   R
L 48–58cm. **W** 100cm. Sexes identical. **N** 0–4,000m. **C & S** <2,000m. **A** <300m. **ID** Similar to Cattle Egret and juvenile Little Blue Heron. Adult has black bill with bare yellow skin, black legs and yellow feet. Juvenile has black bill with pale yellow skin on face and greenish or yellowish legs soiled black. **Habitat** All types of wetlands and coasts. **Voice** Generally silent, except when alarmed and when breeding. Most frequently heard is short, hoarse and rough-sounding *Kaok … Kaok*. **Where to see** Lluta wetland (Arica y Parinacota Region), Elqui River (Coquimbo Region), Lampa/Batuco system (Metropolitan Region), etc.

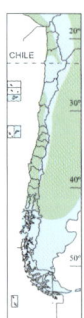

Cattle Egret
Bubulcus ibis   R
L 45–51cm. **W** 94cm. Sexes identical. **N** 0–2,400m. **C & S** <1,000m. **A** <300m. **ID** Similar to Snowy Egret and juvenile Little Blue Heron. Yellow-orange bill and grey, green or yellowish legs and feet. Note posture. **Habitat** Wetlands, agricultural fields or pastures, often associated with livestock or ploughs. **Voice** In escape flight gives a short croak of alarm. Very vocal at communal roosts and in colonies. **Where to see** Laguna El Peral (Valparaíso Region), agricultural fields, landfills. [Alt. Western Cattle Egret]

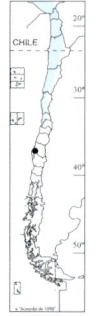

Little Blue Heron
Egretta caerulea   M/R?
L 56cm. **W** 104cm. Sexes identical. **N** 0–2,500m. **ID** Adult unmistakable. Juvenile could be confused with Great Egret, Snowy Egret and Cattle Egret, but has grey bill with black tip. Immature plumage has variable bluish-grey mottling on white background. **Habitat** Always near water (coasts, rivers, interior wetlands, etc.). **Voice** Only generally heard in aggressive encounters or in escape flight, when gives a short, harsh *ghaaekk*. **Where to see** Lluta wetland and Arica's rocky coastal areas (Arica y Parinacota Region), Cavancha Peninsula (Tarapacá Region), Elqui River (Coquimbo Region), Humedal El Culebrón (Coquimbo Region).

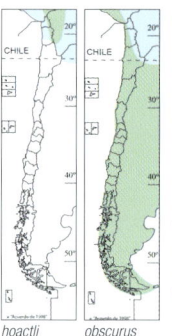

hoactli   *obscurus*

Black-crowned Night Heron
Nycticorax nycticorax   R
L 55–56cm. **W** 112cm. Sexes identical. **N** 0–4,600m. **C, S & A** 0–1,500m. **ID** Unmistakable. Note robust appearance and dark coloration. Juvenile easily confused with juvenile Yellow-crowned Night Heron, which see. **Habitat** Wetlands, fields or coasts. May roost far from water. **Voice** Vocalises little, but its harsh, lugubrious croak (*wock* or *quock*) is often heard in flight. **Where to see** Arica port, Chungará Lake (Arica y Parinacota Region), Rio Maipo wetland (Valparaíso Region), Lampa/Batuco system (Metropolitan Region), Chacao Channel (Los Lagos Region), Punta Arenas (Magallanes Region). **Note** In Arica, it is possible to find both subspecies perched together, always with ssp. *obscurus* predominant in the lowlands and ssp. *hoactli* predominant in the highlands.

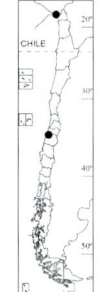

Yellow-crowned Night Heron
Nyctanassa violacea   V
L 55–61cm. **W** 100–110cm. Sexes identical. **N** 0–50m. **ID** Adult unmistakable. Juvenile and immature very similar to Black-crowned Night Heron, but have more erect posture, longer legs and larger bill. In flight, legs project beyond tail. **Habitat** Coastal wetlands. **Voice** Mainly silent, and most vocalisations similar to those of Black-crowned Night Heron, but slightly higher-pitched (*Weeck … Weeck…*). **Where to see** Arica port and Lluta wetland (Arica y Parinacota Region).

PLATE 35: BITTERN, HERONS AND IBISES

Stripe-backed Bittern
Ixobrychus involucris   **R**
L 33cm. Sexes identical. **C & S** <800m. **ID** Unmistakable in range. Small heron, buff coloured and strongly striated, with noticeable dark lines on back. Always perched on the reeds, silent and hidden, following their movement. **Habitat** Closely associated with cattail (*Typha* sp.) and other reed (*Scirpus* sp.) marshes, around lagoons. **Voice** Silent, except when breeding, when low guttural *hak ... hak ... hak ...* is very characteristic, as well as a peculiar, very low-pitched *Booomb... booomb*. **Where to see** Dense reedbeds. El Peral lagoon (Valparaíso Region), Lampa/Batuco system (Metropolitan Region), Petrel lagoon (O'Higgins Region), Putu wetland (Maule Region).

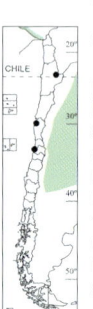

Striated Heron
Butorides striata   **V**
L 36–37cm. Sexes identical. **N, C & S** <300m. **ID** Unmistakable. Compact, with a stockier body than Stripe-backed Bittern. Adult has black cap, grey neck, a blue-grey back and greenish wing feathers with buff fringes. Often exposed or in trees, not hidden in reeds. **Habitat** Wetlands. **Voice** Silent, except in flight or alarm, when gives a high-pitched, shrill *tiuk*, repeated several times. **Where to see** Records in Lluta Valley (Arica y Parinacota Region), Coquimbo area and Elqui River drainage (Coquimbo Region). [Alt. Green-backed Heron]

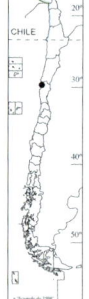

Tricoloured Heron
Egretta tricolor   **M**
L 66cm. **W** 91cm. Sexes identical. **N** 0–30m. **ID** Unmistakable. A slim-bodied heron with a very long neck and dark upperparts. **Habitat** In Chile, only in coastal wetlands. **Voice** Silent, but in escape flight or aggression utters a rough grunt (*gaaah-gaah-gaah*). **Where to see** Lluta wetland (Arica y Parinacota Region).

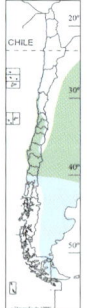

White-faced Ibis
Plegadis chihi   **R**
L 54–56cm. **W** 94cm. Sexes identical. **C & S** <800m. **ID** Unmistakable when perched. In flight could be confused with Black-faced Ibis. Legs extend beyond tail in flight (noticeable even at long range and against the light). **Habitat** Shallow wetlands, waterlogged areas, marshes. Muddy areas with reeds or cattails. Also wet meadows and seasonal wetlands. **Voice** Relatively silent, but calls in contact while feeding and in flight. Relatively noisy around colonies. Harsh, metallic calls, unpleasant to the ear. **Where to see** El Culebrón wetland (Coquimbo Region), Lampa/Batuco system in winter (Metropolitan Region), Budi lake (Araucanía Region)

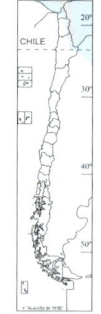

Puna Ibis
Plegadis ridgwayi   **M**
L 56–60cm. Sexes identical. **N** 0–4,000m. **ID** Unmistakable in range. Largely dark ibis. Curved bill and dark body colour are unique in Chilean altiplano. Allopatric with similar White-faced Ibis. **Habitat** High-Andean bogs (bofedales) and lakes. Frequent on coast in far north (Lluta wetland). **Voice** Contact calls, while feeding and especially in flight, are harsh and nasal-sounding (*Quaeck ... Quaeck ...*). **Where to see** Parinacota wetland and Chungará lake (Arica y Parinacota Region), Arabilla lagoon (Tarapacá Region). **Note** Sometimes said to be resident in Chile, but nesting unproven, and considered a visitor here.

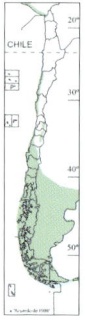

Black-faced Ibis
Theristicus melanopis   **R**
L 74cm. Sexes identical. **N** 0–100m. **C** 0–2,600m. **S & A** <800m. **ID** Unmistakable. In flight could be confused with White-faced Ibis, but legs do not reach beyond tail. **Habitat** Varied environments, but always in open areas. In the north, on coast and islets; in central Chile, recorded in mountains, but from Los Angeles south occurs in lowland fields, around lakes or in natural grasslands. **Voice** Very noisy, calling in flight, from the ground while feeding, at roosts and colonies. Characteristic metallic sounds. **Where to see** Punta de Choros (Coquimbo Region), Farellones (Metropolitan Region), and urban areas from Temuco City (Araucanía Region) to the south.

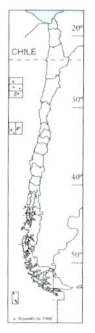

Andean Ibis
Theristicus branickii   **M/V**
L 75cm. Sexes identical. **N** <500m. **ID** Unmistakable. Black in belly begins in thighs, does not have black throat wattle; short legs. Could be confused with Black-faced Ibis. **Habitat** In lowlands, although usually in high Andes outside Chile. **Voice** Silent when foraging, but calls in flight similar to Black-faced Ibis, but shorter and drier, without metallic nuance and even less strident. **Where to see** Restricted to Lluta Valley (Arica y Parinacota Region). **Note** Recent records show Black-faced Ibis and Andean Ibis together in Lluta Valley.

PLATE 36: OYSTERCATCHERS, STILT AND AVOCET

American Oystercatcher
Haematopus palliatus   **R**

L 43cm. Sexes identical. 0–10m. **ID** Similar to Magellanic Oystercatcher (partial overlap). Brown mantle and wings, with black neck, head and chest. **Habitat** Sandy beaches and dunes. Also adjacent islets. Never inland. **Voice** Similar to other oystercatchers. Monosyllabic, high-pitched *peep* in contact by foraging pair. In escape flight, commonly in pairs, gives a shrill-sounding *plee-plee-plee-plee-plee-plee-pleeplee-plee*…. In display, in pairs or groups of three or four, one starts to call and is followed by rest (*fui…fui…fui-fui-fui-fui-fuifui-fui…*). **Where to see** Arica (Arica y Parinacota Region), Caldera (Atacama Region), Rio Maipo wetland (Valparaíso Region), Mataquito River (Maule Region), etc.

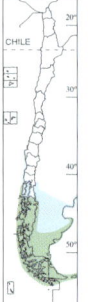

Magellanic Oystercatcher
Haematopus leucopodus   **R**

L 44cm. Sexes identical. 0–500m. **ID** Similar to American Oystercatcher. Mantle and wings black, concolorous with head, neck and breast. **Habitat** Stony beaches, islets and sometimes inland, on grassy prairies and around freshwater lakes. **Voice** Clearly different from other oystercatchers. Powerful and high-pitched. Very strident alarm, given several times, a long monosyllabic *peeeeep* or disyllabic *peee-weep*, becoming even higher-pitched and louder when under threat, before flying away still calling. Territorial (?) calls, given in aggression by two or more pairs, start similarly, but become a metallic trill before returning to typical *peep* calls (*peep-peep-peep-peep-Krrrrrrrrrrrpeepee-peep-peep*). **Where to see** Chiloé Island (Los Lagos Region), road to Fuerte Bulnes, Torres del Paine National Park (Magallanes Region).

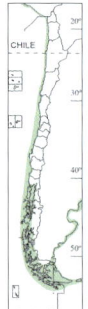

Blackish Oystercatcher
Haematopus ater   **R**

L 52cm. Sexes identical. 0–10m. **ID** Unmistakable. Unique all-blackish oystercatcher (duller in juvenile) with yellow eye and red orbital ring, chisel-shaped red bill and pale pinkish legs. Almost always walking and feeding amongst the rocks. **Habitat** Rocky coasts and small islands nearby. Sometimes on sandy beaches, but never inland. **Voice** Similar to American Oystercatcher. On perceiving a potential threat, gives a monosyllabic, sharp, repeated *pweep*, which serves to alert other pair member, and becomes higher-pitched and louder in alarm, before taking flight. When displaying in pairs, both individuals give similar calls to those just described, followed by a long trill and high-pitched notes (*pweep-pweep-pweeppwep-prrrrrrrrrr-pepepepepe*). **Where to see** Alacrán Peninsula, Arica (Arica y Parinacota Region), Cavancha Peninsula, Iquique (Tarapacá Region), Caldera (Atacama Region), Rio Maipo wetland (Valparaíso Region), Chiloé Island (Los Lagos Region), road to Fuerte Bulnes (Magallanes Region).

Black-necked Stilt
Himantopus mexicanus   **R**

L 42cm. Sexes similar. 0–800m. **ID** Unmistakable. The black-and-white pattern, long pinkish-red legs and needle-like black bill are immediately diagnostic of this species. Juvenile has more blackish feathering on head. **Habitat** Freshwater or semi-brackish wetlands inland or on coast; also sandy beaches. **Voice** Very noisy, in contact and alarm. Pair gives high-pitched monosyllabic calls, at moderate volume and sometimes very slowly, but on perceiving a threat this becomes higher-pitched and louder, and may be repeated by other individuals nearby (*kiep*). When threat becomes more pronounced, the notes are repeated more rapidly (*kiep-kiep-kiep-kiep*), while the bird takes flight. **Where to see** Rio Maipo wetland (Valparaíso Region), Lampa/Batuco system (Metropolitan Region).

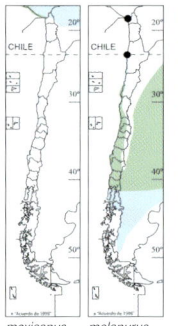

mexicanus   *melanurus*

Andean Avocet
Recurvirostra andina   **R**

L 45cm. Sexes similar. 2,500–4,600m. **ID** Unmistakable, even at a distance. The black-and-white pattern, upturned bill and behaviour of this species are unique in Chilean altiplano. Legs are bluish-grey. Shows white rump in flight. **Habitat** Puna wetlands including lakes, salt flats and high-Andean bogs (bofedales). Occasionally wanders to coastal wetlands. **Voice** Very vocal. Most frequently heard is an alert *kweit, kweit, kweit, kweit*…, also given in escape flight or when mobbing a potential nest predator. In alarm, the notes become higher-pitched and more powerful, *keip-keip-keip-keip*, as the bird takes flight and sometimes circles the potential threat. **Where to see** Chungará lake (Arica y Parinacota Region), Surire Salt Flat (Tarapacá Region), Soncor lagoon (Antofagasta Region).

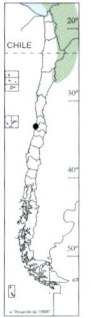

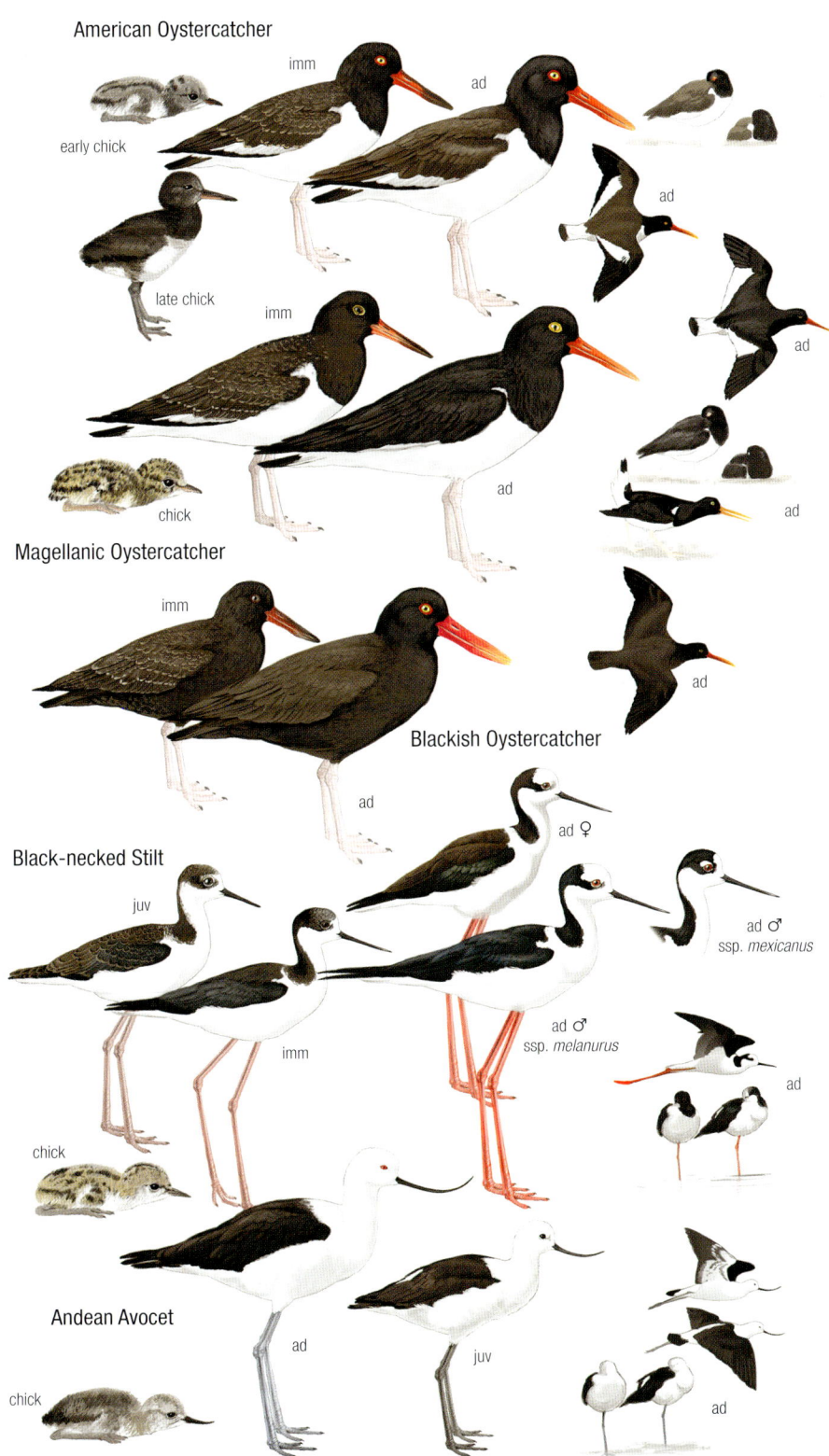

PLATE 37: SANDPIPERS AND ALLIES I

Marbled Godwit
Limosa fedoa   V
L 43–46cm. Sexes identical. 0–10m.
ID Very long bill, curved slightly upwards, and generally ochraceous coloration. **Habitat** Wetlands, beaches, estuaries and muddy bays. **Voice** Silent in Chile. **Where to see** No regular sites. Most likely in September–March.

Hudsonian Godwit
Limosa haemastica   M
L 36–44cm. In non-breeding plumage, sexes identical. In breeding plumage, sexually dimorphic. 0–100m. **ID** Long bill, curved slightly upwards, and generally grey coloration. In all plumages, in flight, note largely black underwing-coverts and black tail with white rump. **Habitat** Wetlands, on muddy banks, estuaries and bays with large intertidal zone. Also inland in far south. **Voice** Most frequently heard is a monosyllabic, short and very sharp *keep* or *qeep*, sometimes repeated several times. **Where to see** Caulín (Los Lagos Region), Lomas Bay (Magallanes Region). Some all year, but most September–March.

Whimbrel
Numenius phaeopus   M
L 42–45cm. Sexes identical. 0–500m.
ID Unmistakable. Decurved bill is diagnostic.
Habitat Marine coasts; rocky areas, sandy beaches, estuaries and marshes. **Voice** Most frequently heard is a *titti-titti-titti-titti-titti-tit* of 5–7 high-pitched notes in flight, sometimes repeated several times, reminiscent of Chilean Tinamou. **Where to see** Virtually any beach. Some all year, but most September–March.

Willet
Tringa semipalmata   M
L 40cm. Sexes identical. 0–10m.
ID Unmistakable. Broad, slightly pointed bill, and note wing pattern in flight. **Habitat** Coastal wetlands, beaches and estuaries; also rocky shores. **Voice** Virtually silent in Chile, giving only weak calls between flock members. **Where to see** Arica (Arica y Parinacota Region), La Serena (Coquimbo Region). Uncommon September–March. **Note** Willet is sometimes considered to be two species. The form in Chile is Western Willet *T. s. inornata*.

Ruddy Turnstone
Arenaria interpres   M
L 21–25cm. Sexes similar. 0–10m.
ID Unmistakable. A stocky, short- and orange-legged wader with a horizontal posture and distinctive black neck-and-breast pattern in all plumages. Commonly seen feeding amongst rocks. **Habitat** Rocky coasts and stony beaches with marine detritus. **Voice** Gives soft vocalisations when in groups, or a harsh sound like weeping in aggressive interactions. In flight a monosyllabic sharp *prii ... prii*. **Where to see** All suitable coasts. Some all year, but most September–March.

Surfbird
Calidris virgata   M
L 23–25cm. Sexes identical. 0–10m.
ID Unmistakable. Robust-bodied, slightly pot-bellied shorebird with whitish eye-ring, yellow legs and a short, straight black bill, with a yellow-orange base of lower mandible in all plumages. Mainly sooty-grey with white rear underparts in non-breeding plumage, but has distinctive rufous in scapulars in breeding attire. Usually seen walking and feeding amongst rocks. **Habitat** Only on rocky coasts, usually among breakers. **Voice** Silent in Chile. **Where to see** Alacrán Peninsula (Arica y Parinacota Region), Cavancha Peninsula (Tarapacá Region), Viña del Mar (Valparaíso Region), etc. Some all year, but large numbers September–March.

Red Phalarope
Phalaropus fulicarius   M
L 18–23cm. In non-breeding plumage, sexes similar. In breeding plumage, sexually dimorphic (not illustrated). 0–10m. **ID** Similar to Red-necked Phalarope, but note grey back, white crown and thicker bill. **Habitat** Pelagic, but occasionally on coast (at lagoons and rivers) in El Niño years. **Voice** Calls among flock members at sea only audible at very close range, mainly monosyllabic croaks. **Where to see** Humboldt Current, between approximately the Arica y Parinacota Region and Los Lagos Region; September–April. [Alt. Grey Phalarope]

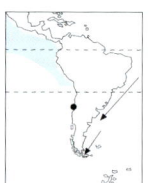

Red-necked Phalarope
Phalaropus lobatus   M
L 15–20cm. In non-breeding plumage, sexes similar. In breeding plumage, sexually dimorphic (not illustrated). 0–10m. **ID** Similar to Red Phalarope. Note lines on upperparts and very sharp, needle-like bill. **Habitat** Pelagic, but occasionally on coast (at lagoons and rivers) in El Niño years. **Voice** Silent in Chile. **Where to see** Humboldt Current in northern Chile (Arica y Parinacota Region). Erratic August–April.

PLATE 38: SANDPIPERS AND ALLIES II

Greater Yellowlegs
Tringa melanoleuca **M**
L 34–36cm. Sexes identical. 0–4,600m.
ID Similar to Lesser Yellowlegs and Stilt Sandpiper. Long legs, neck and bill (which is curved slightly upwards), and secondaries speckled white, visible in flight. **Habitat** Inland or coastal, permanent or temporary wetlands. **Voice** Shy, calls in escape flight, with the Chilean name, Pitotoy, being onomatopoeic (*piii-to-toy ... piii-to-toy ...*), but also gives a single-note whistle (*fuiiii*). Very similar to Lesser Yellowlegs. **Where to see** All year, mostly January–April. Lluta wetland (Arica y Parinacota Region), Lampa/Batuco system (Metropolitan Region), Rio Maipo wetland (Valparaíso Region), but almost any wetland.

Lesser Yellowlegs
Tringa flavipes **M**
L 24–27cm. Sexes identical. 0–4,600m.
ID Similar to Greater Yellowlegs and Stilt Sandpiper. Legs and neck both relatively short. Thin, noticeably fine bill. Secondaries unspotted, visible in flight. **Habitat** Prefers freshwater wetlands, inland or coastal. **Voice** Frequently calls when alert (*pi-pi-to-toy ... pi-pi-to-toy ...*) or in alarm, most about to fly (*pew-pew-pew-pew ... pew-pew-pew-pew ...*). Very similar to Greater Yellowlegs. **Where to see** All year, mostly September to April. Lluta wetland (Arica y Parinacota Region), Lampa/Batuco system (Metropolitan Region), Rio Maipo wetland (Valparaíso Region).

Stilt Sandpiper
Calidris himantopus **M**
L 22cm. Sexes identical. 0–4,200m.
ID Similar to Lesser Yellowlegs and Greater Yellowlegs. Long, slightly decurved bill, small eyes and whitish supercilia that almost reach nape. **Habitat** Varied coastal and inland wetlands. **Voice** High-pitched flight call. **Where to see** Lluta wetland (Arica y Parinacota Region), where relatively frequent. Present September–May.

Wilson's Phalarope
Steganopus tricolor **M**
L 22–24cm. In non-breeding plumage, sexes similar. In breeding plumage, sexually dimorphic. 0–4,600m. **ID** Unmistakable. Upperparts grey with a long, very thin bill. Swims in circles. **Habitat** Inland wetlands, including saline waters; sometimes at coastal wetlands. **Voice** Generally quiet in Chile, calling only in flight or when interacting in groups. **Where to see** Chungará lake (Arica y Parinacota Region), Atacama Salt Flat (Antofagasta Region), Tres Puentes wetland (Magallanes Region).

Spotted Sandpiper
Actitis macularius **M**
L 17–20cm. Sexes identical. 0–10m.
ID Compare Solitary Sandpiper. White of breast reaches around bend of wing. Very characteristic walk, teetering or wobbling. **Habitat** Rocky beaches and mouths of rivers, streams, lakes and lagoons. **Voice** Only usually heard in flight, a high-pitched whistle of 2–3 syllables. **Where to see** Lluta wetland (Arica y Parinacota Region).

Pectoral Sandpiper
Calidris melanotos **M**
L 19–24cm. Sexes identical. 0–4,200m.
ID Unmistakable. Compact, with streaked breast, yellow legs and a medium-length bill that is hardly decurved. **Habitat** Coastal and inland wetlands, including salt lakes. **Voice** Heard in flight or when interacting with others, giving a sharp twitter, repeated several times. **Where to see** Lluta wetland (Arica y Parinacota Region), Atacama Salt Flat (Antofagasta Region).

Short-billed Dowitcher
Limnodromus griseus **V**
L 26–28cm. Sexes identical. 0–50m. **ID** Resembles Magellanic Snipe (Plate 42), but has somewhat longer legs, less cryptic feathering and different behaviour. Also resembles Stilt Sandpiper in non-breeding plumage, but Short-billed Dowitcher is longer and normally retracts neck, which appears shorter. **Habitat** Shallow coastal wetlands and marshes. **Voice** Generally quiet in Chile. Three-note whistle in flight. **Where to see** Only accidental in Chile.

Solitary Sandpiper
Tringa solitaria **V**
L 24–27cm. Sexes identical. 0–4,600m.
ID Similar to Spotted Sandpiper, but has obvious white eye-ring and much darker upperparts, while white of upperparts does not project around bend of wing. **Habitat** Prefers freshwater wetlands, inland or coastal. **Voice** A shrill whistle of 2–4 syllables, sometimes repeated (*twet-twet ... twet-twet-twet-twet ...*). Like others of genus, calls when alert or in flight. **Where to see** Accidental. Records October–June.

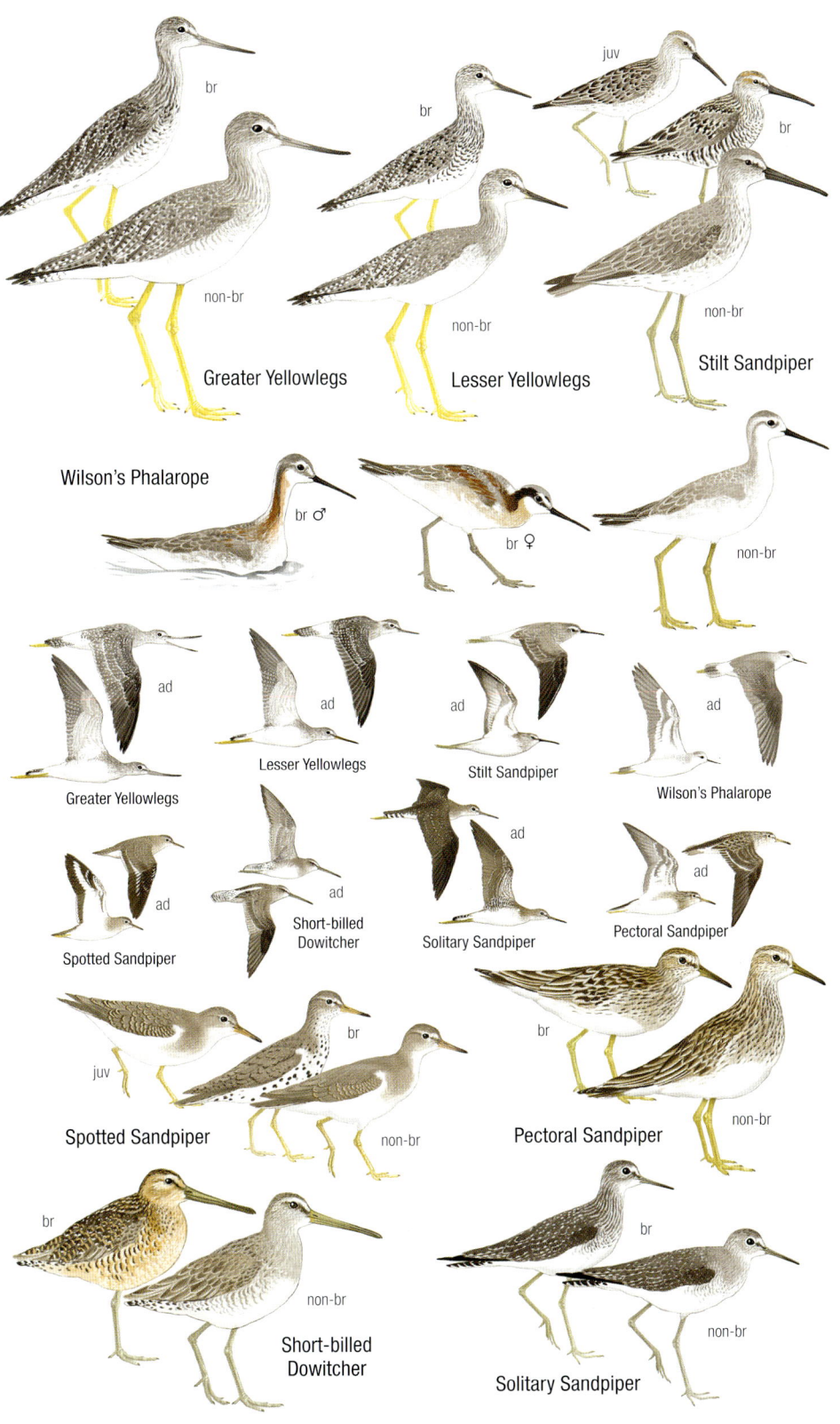

PLATE 39: SANDPIPERS III

It is useful to separate smaller sandpipers 'peeps' into three categories: those with long wings (Baird's and White-rumped Sandpipers); wings that do reach beyond the tail (Semipalmated, Western and Least Sandpipers); and the unmistakable Sanderling.

Baird's Sandpiper
Calidris bairdii　　　　　　　　　　　M
L 14–18cm. Sexes identical. 0–4,600m. **ID** Similar to White-rumped Sandpiper. Brown above, with a straight fine-tipped bill, and dark rump. No spots or streaks on flanks, which are white. Perched, wings project beyond the tail. **Habitat** Coastal and inland wetlands, even at high altitudes, especially in the north (more than 4,000m). **Voice** Calls frequently, mainly a rough single note in contact (*preeep*) or, repeated, in warning (*preeep-preeep-preeep...*), or finally, repeated many times, when flying off. Another monosyllabic call is strident and louder (*tweeeet*), but is heard less, though sometimes repeated. **Where to see** Chungará lake (Arica y Parinacota Region), Atacama Salt Flat (Antofagasta Region), El Yeso Reservoir (Metropolitan Region), etc. Year-round, especially austral spring and summer.

White-rumped Sandpiper
Calidris fuscicollis　　　　　　　　　M
L 16–18cm. Sexes identical. 0–300m. **ID** Similar to Baird's Sandpiper. Greyish-brown above, with thick-based and slightly decurved bill, and pale spot at base. A few dark spots or streaks on flanks. White uppertail-coverts (hence name). Wings project beyond the tail at rest. **Habitat** Coastal and inland wetlands. **Voice** Often not heard due to the very windy environments it inhabits in Chile. Most frequently heard is a sharp, repeated *zip ... zip ... zip ... zip ...*, heard in flight. **Where to see** Tres Puentes wetland (Magallanes Region), Los Cisnes lake and around Porvenir (Magallanes Region). September–April.

Western Sandpiper
Calidris mauri　　　　　　　　　　　V
L 14–17cm. Sexes identical. 0–10m. **ID** Very similar to Semipalmated Sandpiper (both have black legs and small webs between toes), but looks proportionately larger-headed, has shorter neck with a more 'elongated' grey-and-white body, and appears heavy-chested. Long bill is slightly curved, with a fine tip. **Habitat** Mainly on sandy coasts. Accidental inland. **Voice** No information from Chile. **Where to see** Probably under-recorded in Chile. Lluta wetland (Arica y Parinacota Region), but might be found at any estuary. Records all year, especially November–March.

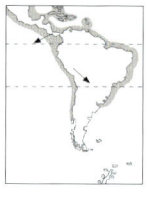

Semipalmated Sandpiper
Calidris pusilla　　　　　　　　　　　M
L 13–15cm. Sexes identical. 0–10m. **ID** Most liable to confusion with Western Sandpiper (which see). Head and whole body appears better proportioned, with a slightly long neck, more rounded body, and straight and robust bill. Less heavy-chested. **Habitat** Mainly on coasts or estuaries. Rare inland. **Voice** A soft and repeated monosyllable, reminiscent of other *Calidris* sandpipers, given in flight (*priii-prii-prii*). **Where to see** Lluta wetland (Arica y Parinacota Region), where relatively frequent. Perhaps under-recorded.

Least Sandpiper
Calidris minutilla　　　　　　　　　　V
L 13–15cm. Sexes identical. 0–10m. **ID** Wings do not reach beyond tail at rest. Smallest sandpiper, with compact body, short neck, short and very thin bill (fine, pointed and with a light curve at the tip) and yellow/greenish legs (can be muddy). Tertials very long, covering primaries. **Habitat** Always at estuaries in Chile. **Voice** No information from Chile. **Where to see** Most frequent at Lluta wetland (Arica y Parinacota Region); records August–April.

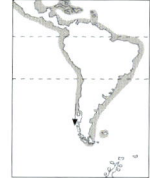

Sanderling
Calidris alba　　　　　　　　　　　　M
L 17–21cm. Sexes identical. 0–10m. **ID** Unmistakable. Short, straight and stubby black bill, mainly white face. Dark shoulder patch not always visible. Juvenile like non-breeding adult, but upperparts and tertials are dark with bold white fringes. Also has streaked crown. **Habitat** Sandy beaches. Rare inland. **Voice** Low-intensity calls given by flock members. In flight occasionally a high-pitched monosyllabic *wick ... wick*. **Where to see** Arica (Arica y Parinacota Region), La Serena (Coquimbo Region), Santo Domingo (Valparaíso Region), Llico (O'Higgins Region), etc. Sandy coasts. Records all year, especially August–May.

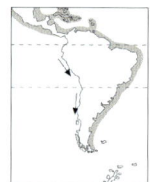

Red Knot
Calidris canutus　　　　　　　　　　M
L 26cm. Sexes identical. 0–10m. **ID** Vaguely recalls Sanderling, but larger and more upright. Robust, rounded body, with a comparatively small head. **Habitat** Sandy beaches with a large intertidal zone. Very rare at inland wetlands. **Voice** Rarely heard in Chile, a high-pitched series of 2–3 syllables (*wek-twek ... twek-twek-twek ...*). **Where to see** Caulín (Los Lagos Region), Lomas Bay (Magallanes Region). Accidental elsewhere. Year-round, especially September–January.

PLATE 40: PLOVERS

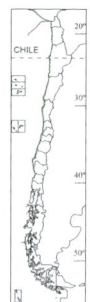

Snowy Plover
Charadrius nivosus   R
L 15cm. Sexes identical. 0–10m. **ID** Similar to Collared Plover. Both adult and immature have breast-side patches. **Habitat** Sandy beaches. It may be in lagoons or nearby estuaries. **Voice** Most commonly heard in escape flight, giving a monosyllabic *peep* that is repeated. **Where to see** Lluta wetland (Arica y Parinacota Region), Santo Domingo (Valparaíso Region), Llico beach (Maule Region).

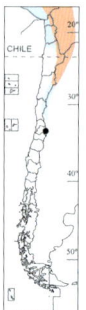

Puna Plover
Charadrius alticola   R
L 18cm. Sexes identical. >2,300m. **ID** On coast, where only accidental, could be confused with Snowy and Semipalmated Plovers. Adult has chestnut crown and nape, in juvenile brown, and barely evident breast-side patches. **Habitat** Puna wetlands such as lakes, lagoons or salt flats, especially muddy and saline areas. **Voice** Calls infrequently, mainly sharp monosyllables (*pioup*) in alarm, sometimes repeated in flight, but becomes more elaborate in territory defence or aggressive interactions (*pioup-pioup-pi-pi-pi-pioup-pi-pipipipipi* ...). **Where to see** Chungará lake and Surire Salt Flat (Arica y Parinacota Region), Puquios (Tarapacá Region), Soncor lagoon (Antofagasta Region).

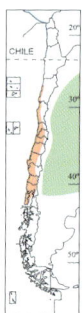

Collared Plover
Charadrius collaris   R
L 16cm. Sexes identical. 0–500m. **ID** Similar to Snowy and Semipalmated Plovers. Adult has black breast-band, brown in immature. **Habitat** Estuaries. Also inland wetlands, sometimes temporary and small, where even observed nesting. **Voice** In general, sharp and metallic. Long *prrep-prrp-prrp-prrp-prrp-prrp-pipipipipip* ... in group disputes (territorial?), with much commotion and displays of aggression among adults. Probably most heard when flying off (*prrip ... ..prrip ... prrip ...*). **Where to see** Elqui River (Coquimbo Region), Rio Maipo wetland (Valparaíso Region), Lampa/Batuco system (Metropolitan Region), Mataquito River (Maule Region).

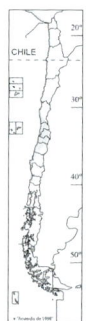

Semipalmated Plover
Charadrius semipalmatus   M
L 17cm. Sexes identical. 0–10m. **ID** Similar to Collared Plover. Adult has bicoloured bill, and all ages have a white hindcollar. **Habitat** Coastal wetlands. Prefers sandy beaches or estuaries. **Voice** Mostly heard in flight, a high-pitched disyllabic, high-pitched *tchiu-wheet*, with accent on second note. **Where to see** Lluta wetland (Arica y Parinacota Region), Pullally Saltworks (Valparaíso Region).

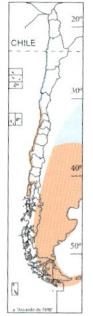

Two-banded Plover
Charadrius falklandicus   R
L 19cm. Sexes identical. 0–300m. **ID** Unmistakable. Only small plover with two complete black breast-bands. **Habitat** Sandy, stony or muddy beaches. Also inland waters. In Patagonia, sometimes meadows or steppe far from water. **Voice** Calls little, but utters a short sharp *puit* from the ground, becoming higher-pitched and given more repeatedly in alarm. Another call appears to be used in aggression or territory defence, a hoarse, rasping *tchk-tchk-tchk- tchk- tchkkrrrrrt* during interactions between several individuals. **Where to see** Pullally Saltworks (Valparaíso Region), Budi lake (Araucanía Region), Otway Sound (Magallanes Region), Los Cisnes lake (Magallanes Region).

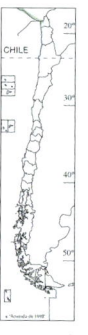

Killdeer
Charadrius vociferus   R
L 23–25cm. Sexes identical. 0–300m. **ID** Unmistakable. Readily identified from its calls. **Habitat** Coastal wetlands, sandy beaches or estuaries, fields and meadows up to 10km inland near Arica. **Voice** Very noisy. A repeated plaintive whistle (*tweeeet*) greets any intruder, and betrays its presence more readily than the bird's cryptic plumage, especially as it is delivered while the bird crouches on the ground. In response to threat, the same whistle is repeated 3–4 times, or many more in flight, *tweeeet -tweeeet- tweeeet - tweeeet... tweetetetetete*. **Where to see** Lluta wetland and Lluta Valley (Arica y Parinacota Region).

PLATE 41: DOTTERELS, SANDPIPER-PLOVER, LAPWINGS AND MAGELLANIC PLOVER

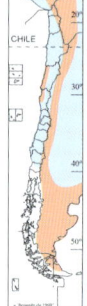

Rufous-chested Dotterel
Charadrius modestus   R
L 22cm. Sexes identical. <1,000m. **ID** Unmistakable in breeding plumage or transition. The rufous-reddish chest is unique and diagnostic of this species. Non-breeding birds change their colourful plumage to brown, but the whitish ventral parts, throat and parts of the diadem remain the same. These changes, and the general shape, are in someway diagnostic of this species. The same happens in juveniles. **Habitat** Variety of wetlands: rocky beaches, inland waters, steppe and Andean meadows in far south (when breeding). Flat coasts, sandy beaches or estuaries in the north (non-breeding). **Voice** Most frequently heard in escape flight is a high-pitched *piu-pi-pi*. Less often heard is a call like stones being repeatedly knocked together, eventually terminating in a toy-like sound (*tik-tik-tik-tik-tik-tik-wrrrrrr* ...), given in flight and from the ground, either during territory disputes or aggression. **Where to see** Road to Fuerte Bulnes and Los Cisnes lake (Magallanes Region). In winter, on beaches and at estuaries in central to northern Chile.

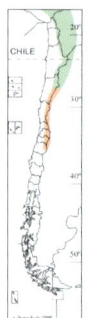

Tawny-throated Dotterel
Oreopholus ruficollis   R
L 25cm. Sexes identical. 0–4,500m. **ID** Unmistakable. Noticeable black spot on belly. Some are more tawny, others greyish, but both have belly spot, even juveniles. **Habitat** Open country, usually very stony and arid, with only sparse vegetation. High-Andean and Patagonian steppes, sometimes near the sea. **Voice** Generally silent, but in flight or social interactions may utter a monosyllabic trill, high-pitched and tremulous, and repeated by one or more birds (*twerrrrrrr*...), apparently in contact among flock members. **Where to see** Suriplaza and Surire Salt Flats (Arica y Parinacota Region), Llanos de Huentelauquen (Coquimbo Region), Pali Ayke National Park (Magallanes Region).

Diademed Sandpiper-Plover
Phegornis mitchellii   R
L 19cm. Sexes identical. >2,400m. **ID** Unmistakable as adult and juvenile. In adults, the combination of dark head, white diadem and rufous nape is unique and diagnostic of this species on the high Andes. In juveniles, the general shape and the brown rufous-cinnamon barred underparts are also diagnostic of this species. Very cryptic. **Habitat** Wetlands (usually bogs) above 2,400m in central Chile, and up to 5,000m in Puna zone, always with some mud. **Voice** Alert call, given in response to potential danger, is a slightly melancholic monosyllable (*wheeeuu* ...). In alarm, especially if there is danger to the chicks, but also in flight, gives a short, sharp *pëw*, repeated sporadically, until the danger passes. Also a very rough grunt, used in aggression, apparently to expel another adult from territory, or between a pair (*gggggggrgrgggggggtgtr*), accompanied by fluffing out the flank feathers, stretching upwards, and making short runs with quick and exaggerated short steps; also given by female seeking to draw attention away from nest. **Where to see** Parinacota (Arica y Parinacota Region), Puxsa Salt Flats (Antofagasta Region), Juncal Andino Park (Valparaíso Region), El Yeso Valley, La Engorda Valley and Colina springs (Metropolitan Region).

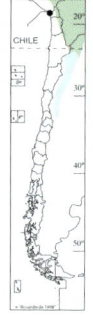

Andean Lapwing
Vanellus resplendens   R
L 32cm. Sexes identical. >4,000m. **ID** Unmistakable and conspicuous. It is the biggest and loudest plover in the environment it inhabits. It has a grey chest, neck and head, with big red eyes and pinkish bare parts. **Habitat** Highland wetlands and open areas. Exceptionally on coast, usually in winter. **Voice** Loud calls, similar to Southern Lapwing, but alarm differs from latter (*kreep- kreep-kreep-ki-ki-ki-ki-kep-kep-kep-kep-kep*). **Where to see** Chungará lake, Guallatire and Parköaylla bofedales (Arica y Parinacota Region).

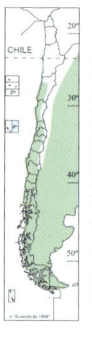

Southern Lapwing
Vanellus chilensis   R
L 36cm. Sexes identical. 0–3,000m. **ID** Unmistakable and conspicuous. Large, and the loudest plover. Very strikingly patterned, with black foreface, foreneck and breast, partially outlined in white, and otherwise grey head and neck, plus bronzy-green shoulders. **Habitat** Pastures, meadows and open country. Coasts. **Voice** Very noisy in response to presence of danger, especially if it has eggs or chicks (*kiew ... kiew ... kiew*). Strident and unmistakable. In flight gives sharp *telëw-telëw-telëw*. **Where to see** Very common. Even in parks and green areas within large cities.

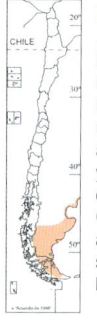

Magellanic Plover
Pluvianellus socialis   R
L 20cm. Sexes identical. 0–300m. **ID** Unmistakable. Recalls a pigeon at distance. Somewhat dove-like shorebird of far south; largely grey with white underparts and tiny bill. Adult has brown breast and pink legs; juvenile is greyish on breast and has yellow legs. **Habitat** Inland lakes or lagoons of fresh or saline water. Also stony or sandy beaches. **Voice** Occasionally gives a monosyllabic, powerful *whééuuu* at intervals. Juvenile higher-pitched and weaker, sometimes disyllabic. **Where to see** Los Cisnes lake, Porvenir (Magallanes Region).

PLATE 42: PLOVERS, SNIPES AND PAINTED-SNIPE

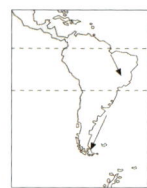

Grey Plover
Pluvialis squatarola **M**
L 29cm. Sexes identical. 0–100m.
ID Similar to American Golden Plover. Note black on axillaries in flight. Also relatively larger, thick and straight bill. Unique in *Pluvialis* in having small hind toe. Never has yellow feathers above (except in early juvenile plumage, not seen in Chile).
Habitat Coastal wetlands. Prefers sandy beaches or estuaries. **Voice** Most commonly heard is a long, high-pitched whistle (*tiiw-iiiit*) given from the ground, or a shorter version in flight. **Where to see** Lluta wetland (Arica y Parinacota Region), Bahía Inglesa (Atacama Region), Rio Maipo wetland (Valparaíso Region). [Alt. Black-bellied Plover]

American Golden Plover
Pluvialis dominica **M**
L 26cm. Sexes identical. 0–4,200m.
ID Similar to previous species. No black in axillaries, but can show some yellow feathers above. Bill is sharp, and slightly triangular. No hind toe, like all *Pluvialis* sp. (except Grey Plover). **Habitat** Coasts and high-Andean wetlands in far north. Lagoons and salt flats. **Voice** In alarm or flight gives two sharp syllables. **Where to see** Most likely October–June. Records Arica and Surire Salt Flat (Arica y Parinacota Region), Atacama Salt Flat (Antofagasta Region), Tierra del Fuego, near Porvenir (Magallanes Region).

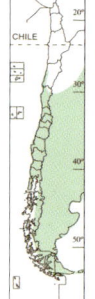

Magellanic Snipe
Gallinago magellanica **R**
L 22–30cm. Sexes identical. **C** 0–3,000m. **S & A** 0–1,000m. **ID** Similar to Puna Snipe (mainly allopatric). In flight recalls South American Painted-snipe and Grey-breasted Seedsnipe. Legs project beyond tail in flight, which is zigzagging. Legs greenish or yellow. **Habitat** Wet, muddy areas to 3,000m. Sometimes far from flooded areas in far south. **Voice** Usually given from the ground is a very rhythmic *kekék-kekék-kekék*, or a rather different *kek-kek-kek-kek-kek*-... In high-speed flight, especially at dusk and dawn, sometimes also at night, splays the tail allowing wind to rush through it, producing a winnowing sound, as the bird either describes a figure of 8 or an ellipse c.8–20m up, before plummeting to the ground. **Where to see** Lampa/Batuco system (Metropolitan Region) and Rio Maipo wetland (Valparaíso Region), Tres Puentes wetland, Bahía Azul road to Porvenir and Torres del Paine National Park (Magallanes Region). **Note** Formerly considered to be a subspecies of South American Snipe *G. paraguaiae*.

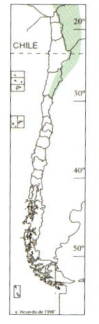

Puna Snipe
Gallinago andina **R**
L 22–25cm. Sexes identical. 2,000–4,600m. **ID** Similar to Magellanic Snipe (mainly allopatric). In flight recalls Grey-breasted Seedsnipe. Short legs are always yellow, and do not project beyond tail in flight. **Habitat** Altiplano wetlands, e.g. high-Andean bogs (bofedales), lakes, salt flats and rivers. **Voice** From the ground gives a repeated *kiek-kiek-kiek-kiek-kiek* ..., which is faster than Magellanic Snipe. Also a high-pitched call in escape flight (*kekek-kekek -kekek* ...). Like Magellanic Snipe, in display, especially at dusk and dawn, sometimes also at night, splays the tail allowing wind to rush through it, producing a winnowing sound, as the bird flies up and down c.8–20m up, before plummeting at speed to the ground. **Where to see** Chungará lake and Parinacota bofedales (Arica y Parinacota Region), Quepiaco wetland (Antofagasta Region).

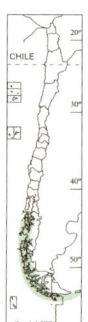

Fuegian Snipe
Gallinago stricklandii **R**
L 29–36cm. Sexes identical. 0–10m. **ID** The largest snipe in Chile. Stocky with a long bill and slightly drooping tip, and tail concolorous with mantle (not orange). **Habitat** Muddy areas, wetlands, Patagonian bogs, high-elevation grasslands. **Voice** Like other snipes in Chile, calls from the ground, in alarm and displays in flight, including at night, using tail feathers. **Where to see** Wollaston Islands and Almirantazgo Sound (Magallanes Region).

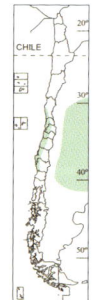

South American Painted-snipe
Nycticryphes semicollaris **R**
L 22cm. Sexes identical. **C & S** 0–600m. **ID** Unmistakable at rest. In flight, similar to Magellanic Snipe, but note curved trajectory, in silence and usually with legs dangling. Curved bill is diagnostic if seen. **Habitat** Flooded, well-vegetated marshes. Also reported at estuaries, in rivers or irrigation canals. **Voice** Very rarely heard, even when escaping from a threat. **Where to see** Lampa/Batuco system (Metropolitan Region), Putú wetland (Maule Region).

PLATE 43: UPLAND SANDPIPER, THICK-KNEE, SEEDSNIPES, QUAIL AND PHEASANT

Upland Sandpiper
Bartramia longicauda **V/M?**
L 28–32cm. Sexes identical. 0–3,500m.
ID Unmistakable. Very distinctively shaped wader. Usually upright posture, with a small head and long tail (for a sandpiper). When perched, it typically leaves its wings open for a moment before folding them. **Habitat** Arid and semi-arid areas. Also high-Andean grasslands. **Voice** Not heard in Chile, but can give a persistent high, two- or three-note *Kip-ip-ip-ip... Kip-ip-ip....* **Where to see** Records concentrated in Antofagasta Region, apparently on migration to/from Argentina.

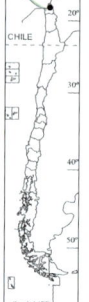

Peruvian Thick-knee
Burhinus superciliaris **R**
L 38–40cm. Sexes identical. 0–500m. **ID** Unmistakable. Very large and conspicuous yellow eyes. Very cryptic plumage; generally it is found in places where its coloration resembles the terrain. It moves very slowly until it is discovered, escaping by walking; rarely flies to flee. **Habitat** Abandoned fields, arid areas with scrub, corn stubble and sandy riverbanks. **Voice** Rather silent by day, but noisy at night, both from the ground and in flight, and sometimes in noisy flocks. Typical is a multisyllabic *werekeke- wereke- werekeke keke- werekeke....* **Where to see** Lluta Valley, Azapa Valley and Chaca Valley (Arica y Parinacota Region).

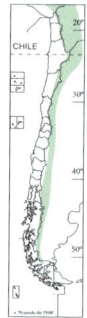

Rufous-bellied Seedsnipe
Attagis gayi **R**
L 27–30cm. Sexes identical. **N** >4,000m. **C & S** >3,000m. **ID** Extremely cryptic plumage, it can disappear in front of the observer if it chooses to stand still. Its general appearance resembles a pigeon walking at a distance. All-rufous underparts, as befit its name. **Habitat** In the north, in high-Andean bogs (bofedales) and near other wetlands. In central Chile, prefers rocky slopes near snowline and streams, but will visit fertile valleys and wet gorges to feed. **Voice** Monosyllabic alarm when threatened, although prefers to remain crouched, or to walk discreetly away from threat. Nevertheless, flock may give *quit-quit-quit-quit*, especially in flight, in a somewhat discordant, metallic chorus. **Where to see** Las Cuevas, Lauca National Park and Surire (Arica y Parinacota Region), Puxsa Salt Flat (Antofagasta Region), El Plomo canyon (Metropolitan Region).

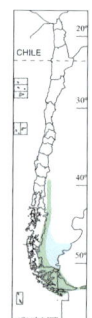

White-bellied Seedsnipe
Attagis malouinus **R**
L 25–29cm. Sexes identical. **A** 650–2,000m. **ID** Unmistakable. Large seedsnipe with white belly. **Habitat** Mountainous areas at treeline or snowline. Rocky slopes with *Empetrum rubrum* and *Bolax gumifera*. In winter descends slightly, to humid ravines and open steppe. **Voice** Rarely calls except when threatened, when usually walks away giving short, high-pitched *pweet, pweet, pweet* notes or a continuous metallic chirping *prrrrrrrrrt*; very similar to Rufous-bellied Seedsnipe. **Where to see** Torres del Paine National Park, Sierra Baguales (Magallanes Region).

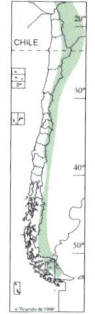

Grey-breasted Seedsnipe
Thinocorus orbignyianus **R**
L 20–23cm. Sexually dimorphic. **N** >4,000m. **C** >2,000m. **S** <1,000m. **ID** Similar to smaller Least Seedsnipe. Male has clean grey breast. **Habitat** Andes; near watercourses or semi-flooded areas with vegetation. In the south, will visit lowland wetlands. **Voice** Territorial song of male, given at any time of day, perhaps especially in twilight, from the ground or in flight, is *p'koy-p'koy-p'koy-p'koy-p'koy*. **Where to see** Las Cuevas (Arica y Parinacota Region), Machuca (Antofagasta Region), Farellones, El Yeso Valley (Metropolitan Region).

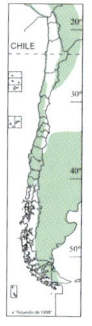

Least Seedsnipe
Thinocorus rumicivorus **R**
L 17–19cm. Sexually dimorphic. **N** >3,000m. **C** >0–2,600m. **A** <1,200m. **ID** Similar to larger Grey-breasted Seedsnipe. Male has vertical black line over central breast (varies from broad to barely marked). Some females show a slight indication of this. **Habitat** Open, arid and stony ground, with low, sparse vegetation. Coastal plateaus. **Voice** Most calls are rather quiet. Performs aerial display lasting 6–10 seconds, comprising a six-syllable low introduction, followed by a melodic phrase repeated with slight variations 4–5 times, ending in 3–4 repetitions of the introduction, but louder and higher-pitched (*hut -hut-hut-huthut-hut-Tchaka-ut-Tchaka-u-ut-ut ... Hut-Hut-Hut-Hut ...*) **Where to see** Lampa/Batuco system (Metropolitan Region), Huentelauquen (Coquimbo Region), Pichilemu (O'Higgins Region), Pali Ayke National Park, Porvenir (Magallanes Region).

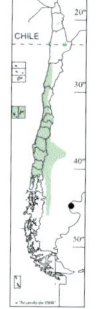

California Quail
Callipepla californica **Ri**
L 25–26cm. Sexually dimorphic. **N** 2,000–2,500m. **C & S** 0–2,000m. **ID** Unmistakable. The peculiar forward-facing topknot, most developed in males, and plumage pattern (especially the 'scales' in chest, belly and flanks) are diagnostic of this species. **Habitat** Common in semi-arid environments and agricultural fields. Scrubby areas. Never in dense forest. **Voice** Sounds are typical of central Chile's fields. Male has powerful territorial song (*Chee-ka ... Chee-ka ... Chee-ka-go*). In alarm, several flock members give a high-pitched *pwet, pwet, pwet, pwet, pwet, pwet*. Also contact calls while foraging. **Where to see** Anywhere in coastal range (Coquimbo Region–O'Higgins Region), Lagunillas road (Metropolitan Region).

Common Pheasant
Phasianus colchicus **Ri**
L 65cm (♂) 55cm (♀). Sexually dimorphic. **A** >1,000m. **ID** Long-tailed gamebird, principally brown in female, but adult male is rather spectacularly plumaged and impossible to confuse with any other bird in Chile. **Habitat** Very local, in open forest, natural meadows or scrub. **Voice** Vocal throughout year, especially males, in courtship and territory defence. Unmistakable call, usually heard during twilight, strident and with a metallic quality (*ko-kok*). In alarm, a persistent *crrrrc, crrrrc, crrrrc, crrrrc*, before taking flight. **Where to see** Mallin Grande. Not mapped. [Alt. Ring-necked Pheasant]

98

PLATE 44: RHEAS AND TINAMOUS

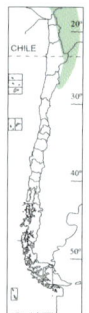

Puna Rhea
Rhea tarapacensis R
L 90–130cm. Sexes identical. **N** >4,000m. **ID** Unmistakable in Chile. The only large, long-necked and flightless bird within its range; generally similar to Lesser Rhea, but ranges quite separate. **Habitat** Steppe and grassland (*Stipa* sp.) with low shrubs. Occasionally high-Andean bogs (bofedales) and salt flats. Recorded at 2,500m near Atacama Salt Flat. **Voice** Mainly silent. Low, hoarse sounds, like grunts or even hisses. In aggression adult produces snoring sounds and bill-snaps. **Where to see** Very local. Near Guallatire (Arica y Parinacota Region), close to Surire (Arica y Parinacota Region).

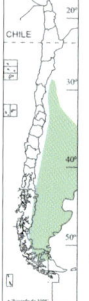

Lesser Rhea
Rhea pennata R
L 95–140cm. Sexes identical. **A** <2,000m. **ID** Unmistakable in Chile. The only large, long-necked and flightless bird within its range. **Habitat** Patagonian steppe, shrub-steppe, grassland. **Voice** Low, hoarse sounds, like grunts or even hisses. Chicks call softly, higher-pitched in alarm. In aggression adult produces snoring sounds and bill-snaps. **Where to see** Widespread in southern Chile; between Punta Arenas and Puerto Natales, road to Sierra Baguales, Pali Ayke National Park (Magallanes Region).

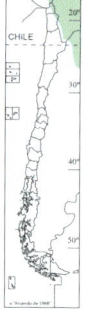

Ornate Tinamou
Nothoprocta ornata R
L 35–37cm. Sexes identical. **N** 2,500–4,300m. **ID** Unmistakable in Chile. Somewhat bushy-crested tinamou, with broad grey shawl on breast and neck, otherwise buffy underparts, relatively plain face and almost ocellated upperparts. **Habitat** Andean slopes with low vegetation, steppe and grassland, low scrub and *Polylepis* forest. **Voice** Most frequently heard, when flushed, is a first shrill whistle, followed by a series of descending whistles (*pweeu ...pweeu... pweeu- pweeu*), like Chilean Tinamou. Territorial song is a short single *fweet*, recalling Austral Thrush. **Where to see** Socoroma, Putre, Chapiquiña (Arica y Parinacota Region).

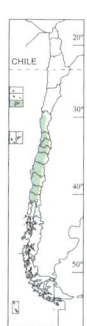

Chilean Tinamou
Nothoprocta perdicaria Re
L 29cm. Sexes identical. **C** 0–2,200m. **ID** Unmistakable. The most widespread tinamou in Chile. Upperparts streaked; grey breast and buffy ventral region. Bushy crest raised frequently when walking. Very cryptic. **Habitat** Grassland, scrub, thickets on arid slopes and steppe. Also open sclerophyll forest, but not dense forests. **Voice** Best known is a powerful territorial whistle, given day and night (*twee – reep*). On flushing a shrill whistle, followed by a series of descending ones (*peeepeepeepee...pweep pweep*). **Where to see** La Campana National Park (Valparaíso Region). Andean slopes, road to Lagunillas (Metropolitan Region).

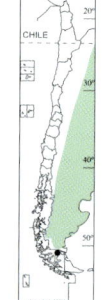

Elegant Crested Tinamou
Eudromia elegans R/V?
L 40–42cm. Sexes identical. **A** <400m. **ID** Similar to Patagonian Tinamou. Long and obvious crest that projects backwards. **Habitat** Patagonian steppe, semi-arid grassland (*Festuca* sp.) and Patagonian low scrub. **Voice** Most often heard is a loud whistle repeated several times (*fu-it ... fu-it... fu-it...*). **Where to see** Very local; Chile Chico and Balmaceda (Aysén Region).

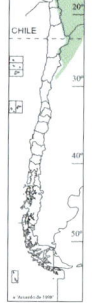

Puna Tinamou
Tinamotis pentlandii R
L 42–46cm. Sexes identical. **N** 3,800–5,000m. **ID** Unmistakable. Large tinamou without a crest; head and neck white and grey, with characteristic pattern. Patagonian Tinamou is wholly allopatric. **Habitat** Highland environments, except flooded areas. Common in low scrub. **Voice** Best known is a chorus of high *keew keew keewa* notes, given by several individuals responding to one another, in early morning. In response to threat, birds walk slowly away, relying on their cryptic plumage, but commonly uttering a high *kew kew kew kew kew kew ...*, which nonetheless betrays their presence. **Where to see** Lauca National Park and road to Surire (Arica y Parinacota Region), road to Miscanti/Miñique and road to Talabre (Antofagasta Region).

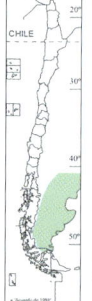

Patagonian Tinamou
Tinamotis ingoufi R/V?
L 39–40cm. Sexes identical. **A** 0–800m. **ID** Similar to Elegant Crested Tinamou. Large tinamou, without crest, head and neck white and grey, with characteristic pattern like Puna Tinamou, but totally allopatric. **Habitat** Patagonian steppe, scrub and shrubby areas, with low vegetation. **Voice** Most often heard is a long *kew-kew-kew- kew- kew-kew- kew-kew...*, given softly and which decreases in intensity, probably with a territorial function, whereas a more strident *kew-la, kew-la, kew-la...* probably signals alarm. **Where to see** No good spots. Most likely between Primera Angostura and Monte Aymond (Magallanes Region); Pali Ayke National Park (Magallanes Region).

PLATE 45: VULTURES

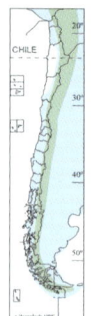

Andean Condor
Vultur gryphus R
L 120cm. **W** 320cm. Sexually dimorphic. 0–5,000m. **ID** Unmistakable perched. At distance in flight could be confused with Turkey and Black Vultures. Huge. All black, except collar and secondaries, which are both white. Seven outermost primaries resemble fingers. Juvenile in flight appears tawny/brownish, gradually changing to black and reaches adult plumage only when c.5 years old. **Habitat** Andes and coastal ranges. Descends to sea level, especially in extreme north and far south of Chile, and even wanders to Santiago city. **Voice** Utters snorts, grunts, and mechanical sounds with the bill, audible only at close range. **Where to see** Lauca National Park (Arica y Parinacota Region), Libertadores Pass (Valparaíso Region), El Pangue landfill (Metropolitan Region), Farellones (Metropolitan Region), Termas del Flaco (O'Higgins Region), Chillán hot springs (Ñuble Region), Isla Riesco/Río Verde, Torres del Paine (Magallanes Region).

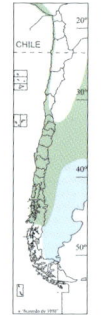

Black Vulture
Coragyps atratus R
L 65cm. **W** 150cm. Sexes similar. 0–2,000m. **ID** Unmistakable perched. At distance in flight could be confused with juvenile and immature Andean Condor and Turkey Vulture. Six outer primaries whitish. Very short tail, beyond which legs protrude slightly in flight. Proceeds with fast fluttering flaps, then soars. When soaring, wings held horizontal to body. **Habitat** All types of environment including coastal cities. **Voice** Mute, but makes mechanical noises with bill, also snorts and grunts, but only audible at close range. **Where to see** La Serena (Coquimbo Region), Cerro Ñielol Natural Monument (Araucanía Region), Chiloé Island (Los Lagos Region).

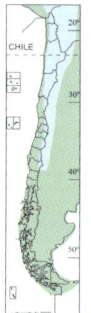

Turkey Vulture
Cathartes aura R
L 71cm. **W** 170cm. Sexes similar. 0–2,000m. **ID** Unmistakable perched. At distance in flight could be confused with juvenile and immature Andean Condor and Black Vulture. Remiges silvery-grey from below. Otherwise blackish-brown in all plumages, except head (which changes). Long tail. Wings held in V when soaring. **Habitat** All manner of environments, from coast to interior, but more abundant in former. Inshore islands. Coastal cities. **Voice** Utters mechanical sounds with bill; also snorts and grunts. Only audible at close range. **Where to see** Coasts between Lluta wetland (Arica y Parinacota Region) south to Punta Arenas (Magallanes Region).

Andean Condor   Turkey Vulture   Black Vulture

PLATE 46: HAWKS I

Field identification of raptors can be complex, given several very similar species, as well as a variety of non-adult plumages and, in some species, different-coloured morphs. Usually seen in flight, against an open sky, making size sometimes difficult to estimate, thereby highlighting the importance of different silhouettes and flight styles. For ease of use, four stages are mentioned: juvenile, or the first plumage in which the young leaves the nest (basic I); early immature, from the second moult cycle (basic II); late immature, from approximately the third (basic III) to fourth moult cycles (basic IV); and, finally, adult plumage (basic definitive). Intermediate plumages are occasionally mentioned to facilitate recognition in the field. It should be understood that there are no discrete steps between stages, merely a continuum, within which individual variation can generate additional confusion.

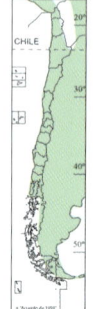

Variable Hawk
Geranoaetus polyosoma   **R**
L 45–51cm (♂) 52–55cm (♀). **W** 113–141cm. Sexually dimorphic. 0–4,500m. Several different plumages pre-adult, and two morphs (pale and dark), each with different male and female plumages, causing considerable confusion, especially for the inexperienced observer. The pale morph is commoner. **ID** In far north, perched adult can be confused with Puna Hawk. In all morphs wings rather rectangular and has white tail, sometimes with slight barring, and broad black subterminal band. Juveniles and immatures, especially paler birds, can be confused with juvenile Harris's Hawk, but wings reach approximately to tip of tail, and belly and flanks are barred, not spotted or streaked. Can also be confused with juvenile Black-chested Buzzard-Eagle, but is relatively smaller and less broad-winged. In the south, dark morph almost indistinguishable from dark juvenile Rufous-tailed Hawk, but note narrower tail bars. **In flight** Well-proportioned silhouette and persistent glides. Sometimes hovers into wind for long periods. Adult always has white tail with broad black subterminal band and slightly longer wings than they are broad. Juvenile and early immature have grey tail with fine dark barring, only acquiring the well-defined subterminal band with age. **Habitat** All types of environment. Mountainous areas (Andes and coastal range), semi-arid areas with low vegetation, sclerophyll and temperate forest (*Nothofagus* sp.), and coastal cliffs. **Voice** Generally silent, only usually heard when breeding. Loud and monotonous, shrill cries, e.g., a repeated *peewp-peewp-peewp-peewp-peewp-peeewp....* **Where to see** Common throughout range. Farellones area, Lo Prado road (Metropolitan Region), Andean foothills and mountainous areas.

juv

imm

ad

PLATE 47: HAWKS II

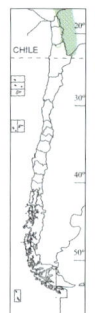

Puna Hawk
Geranoaetus (polyosoma) poecilochrous   R
L 48–51cm (♂) 55–58cm (♀). **W** 133–151cm. Sexually dimorphic. 3,400–5,000m. **ID** Almost indistinguishable from Variable Hawk in all ages and morphs, but their ranges overlap only slightly. A robust hawk (male as large as female Variable Hawk), with very broad wings and a proportionately short tail. **In flight** Well-proportioned silhouette and slow glides. Sometimes hovers against wind. Adult always has white tail with broad black subterminal band and broad wings. Juvenile and early immature have grey tail with fine dark barring, only acquiring subterminal band with age. Wings appear rounded in silhouette, due to relative length of inner primaries and especially secondaries. Tail appears very short, due to the wing shape. **Habitat** Altiplano environments. Some descend through the Loa River canyon almost to sea level, but remain in the vicinity of the river, which crosses a true desert. **Voice** Generally silent, but young birds, especially, vocalise frequently during the breeding season. Differs from Variable Hawk. Most frequently heard, apparently in territory advertisement, is a repeated *kiew-kiew-kiew-kiew*.... **Where to see** Socoroma/Putre/Parinacota (Arica y Parinacota Region), Isluga/Enquelga (Tarapacá Region), Machuca, Loa River canyon (Antofagasta Region). **Note** Despite the existence of publications that consider *G. polyosoma* and *G. poecilochrus* conspecific, the evidence still does not seem irrefutable. The authors have chosen to keep both species separate, due to their differences (morphology, habitat, voice), but doubts remain until a definitive study is published. For this reason *G. (polyosoma) poecilochrus* has been used as the designation for Puna Hawk.

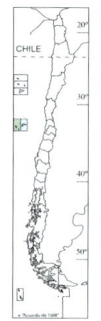

Masafuera Hawk
Geranoaetus polyosoma exsul   R
L 48–54cm. **W** 123–138m. Sexually dimorphic. **ID** Unmistakable. The only large raptor on the Juan Fernández Islands. Lacks obvious reddish mantle of mainland Variable Hawks. **Habitat** Found in all parts of Alejandro Selkirk Island. Rare visitor to other islands on Juan Fernandez archipelago. **Note** Presented in a separate account here, but usually treated as a subspecies of Variable Hawk. [Alt. Juan Fernandez Hawk]

ad ♂
dark morph

early ad ♂
dark morph

ad ♀

PLATE 48: HAWKS III

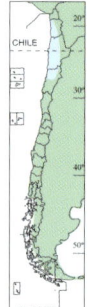

Black-chested Buzzard-Eagle
Geranoaetus melanoleucus   **R**
L 60–76cm. **W** 149–184cm. Sexes similar. **N** 0–3,600m. **C** 0–2,500m. **S & A** 0–1,000m. **ID** Adult unmistakable. Juveniles and some immatures could be confused with juvenile and immature Variable and Puna Hawks. A very robust raptor, relatively large with spotted plumage, and an ochre or dirty white breast. **In flight** Adult in flight resembles an inverted triangle. Very short tail is black. Juvenile different, with wedge-shaped tail, much longer than adult, and has broader wings. Immature has silhouette like juvenile, but a shorter tail. Soaring flight, with few wingbeats. **Habitat** Wide variety of environments. Mainly in mountains and foothills (Andes and coastal range) in northern and central Chile. In the south, occupies prairies associated with forests and cliffs. **Voice** Usually silent, but when breeding gives loud calls when approaching nest and in alarm, uttering a high-pitched *khakhakhakhakhakha*. Young give a high-pitched whistle to attract their parents (*kiew..kiew..kiew…kiew…*). **Where to see** Andean foothills, e.g. road to Farellones and Aguas de Ramón Natural Park (Metropolitan Region), and coastal range, e.g. La Campana National Park (Valparaíso Region), Torres del Paine National Park (Magallanes Region).

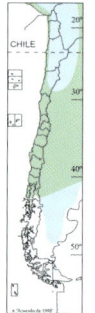

Harris's Hawk
Parabuteo unicinctus   **R**
L 47–57cm. **W** 92–121cm. Sexes similar. 0–1,500m. **ID** Adult unmistakable. Juvenile and immature can be confused with Variable Hawk, but when perched tail is noticeably longer than wings, and belly is spotted or streaked dark, not barred. **In flight** Broad rounded wings, with central secondaries longer than rest. Long tail. **Habitat** Varied environments, but favours arid canyons or small valleys, foothills and scrubland. Also penetrates large cities, especially juveniles in winter. Generally does not reach high elevations, but one record at Putre (3,600m). **Voice** Generally silent, except during breeding season, both adults and juveniles. Immature gives a very high-pitched *whee-whee-whee*…. Most characteristic is a strident, very high-pitched *keeeeeeee*…, often given in flight. **Where to see** Lluta Valley (Arica y Parinacota Region), Lampa/Batuco system, road to Farellones and Mahuida Park (Metropolitan Region), La Campana National Park (Valparaíso Region), coastal range (Coquimbo Region–Maule Region).

PLATE 49: HAWKS IV

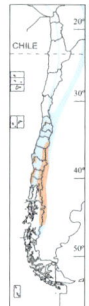

White-throated Hawk
Buteo albigula R
L 47–52cm (♂) 54–57cm (♀). **W** 95cm. Sexes similar. 0–2,500m. **ID** Adult can be confused with juvenile of larger Rufous-tailed Hawk. Note white throat and upper breast without dark spots, dark breast-sides and flanks, and scattered brown elongated spots on belly. Juvenile has pale brown iris; dark brown in adult. **In flight** Medium-sized raptor with short rounded wings, and mid-length tail. Soars in wide circles at great height above ground. Shallow undulating flight, sometimes with dives, in courtship. Stoops into forest at great speed. **Habitat** Mountains with native forest from sea level to c.2,500m, especially forests of *Nothofagus* sp. **Voice** Mostly quiet, except in territory advertisement. The commonest vocalisation is a shrill, long and very high-pitched shriek, repeated in warning (*Tweeeeeiii ... Tweeeeeiii... Tweeeeeiii... Tweeeeeiii... ..*). Also a shorter and high-pitched *Pweu ... Pweu ... Pweu ...*, repeated frequently. **Where to see** Ñilhue bridge (road to Farellones) (Metropolitan Region), La Campana National Park (Valparaíso Region), Termas de Chillán (Ñuble Region), Tolhuaca (Araucanía Region).

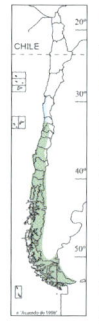

Chilean Hawk
Accipiter chilensis R
L 37cm (♂) 42cm (♀). **W** 58–83cm. Sexes similar. **C** 0–2,000m. **S & A** 0–1,000m. **ID** Unmistakable. A smaller raptor with a long, clearly banded tail but relatively short and rounded wings. **In flight** Silhouette of short, rounded wings and long tail; when hunting, flight fast and agile among trees. **Habitat** Temperate forest (*Nothofagus* sp., *Araucaria araucana*). Sclerophyll forest in central zone. Sometimes observed in open scrub in far south. **Voice** Typically quiet, except when breeding, calling in territorial display flights, alarm and contact at nest, including changeovers between male and female. Most often heard is a high, slightly metallic call with distant resemblance to the call of Ringed Kingfisher (*kak-kak-kak-kak-kak..*). **Where to see** La Campana National Park (Valparaíso Region), Alto Vilches (Maule Region), Termas de Chillán (Ñuble Region), Punta Arenas, including the main square in winter (Magallanes Region). **Note** Some authors treat it as a subspecies of Bicoloured Hawk *A. bicolor*.

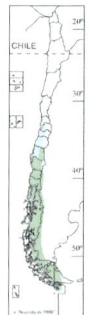

Rufous-tailed Hawk
Buteo ventralis R
L 54–56cm (♂) 57–60cm (♀). **W** 114–139cm. Sexes similar. 0–1,500m. **ID** Similar to Variable and White-throated Hawks, and even Harris's Hawk. All plumages (except dark morph) show dark leading edges (dark patagia) to wing. All plumages (except first year and dark morph) show dark trailing edge to wing. Tail always dark, with obvious bars, rarely with slight reddish tint. Juvenile dark morph almost indistinguishable from Variable Hawk at same age/morph, but has broader tail bars and more obvious yellow iris. In all morphs, yellow iris in juvenile, brown in adult. **In flight** Large, appears robust compared to congeners. Long broad wings. Note underwing pattern. **Habitat** Patagonian-Andean temperate forest, transition to shrub-steppe and mountain forests. **Voice** Only really vocal when breeding. A high-pitched call of four or five syllables (*keee-keee-keee-keee-keeeu*), also a higher *kweu* or longer *kweeuu* by pair in vicinity of nest. **Where to see** Cerro Ñielol Natural Monument, Cordillera de Nahuelbuta National Park (Araucanía Region).

PLATE 50: HARRIER, KITE AND OSPREY

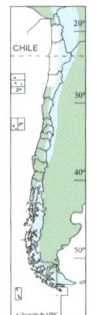

Cinereous Harrier
Circus cinereus   R
L 40cm (♂) 50cm (♀). **W** 90–115cm. Sexually dimorphic. **C** 0–2,500m. **ID** Unmistakable. Facial disc recalls a Barn Owl, and always shows an obvious white rump. **In flight** Long wings and tail. Both sexes and all ages show white rump. Usually flies low, swaying from side to side, checking the ground with wide changes of directions. Glides strongly, the wings held in a slight V. **Habitat** Mostly associated with wetlands. Females (both sexes when breeding) generally prefer wetlands with abundant vegetation, low ground and grassland. In non-breeding season males can be found far from water, in meadows, foothills and even in high-Andean fields. **Voice** Only vocal when nesting. Male gives a repeated, monosyllabic *kiew … kiew … kiew …*, female a faster *kiek-kiek-kiek-kiek-kiek-kiek*, which in alarm becomes even higher and more rapid (*kiekiekiekiekiekiekie*). Pairs in flight may give surprisingly soft, somewhat twittering notes. **Where to see** Huasco River (Atacama Region), Punta Teatinos (Coquimbo Region), Lampa/Batuco system (Metropolitan Region), road to Guabún (Los Lagos Region), Torres del Paine National Park (Magallanes Region).

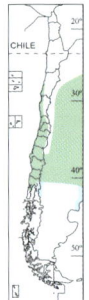

White-tailed Kite
Elanus leucurus   R
L 37–40cm. **W** 100cm. Sexes similar. 0–1,000m. **ID** Unmistakable. The palest of the Chilean raptors. Perched adult is white from the front and pale grey in dorsal view. Scapulars and lesser coverts are both black, and very noticeable, contrasting with dorsal grey. Red eyes. **In flight** Long pointed wings and long tail. Buoyant soaring flight. Sometimes hovers with rapid wingbeats and spread tail. Can recall a gull in flight. **Habitat** Prairies, pastures, cleared land, agricultural fields, wet areas and coastal dunes. Requires some trees, but can adapt to introduced *Eucalyptus* sp. **Voice** Vocal only during breeding season (*kewak..kewak…*). Chick and juvenile give calls reminiscent of Barn Owl when begging, while adult utters a short whistle. **Where to see** Cultivated plains near Melipilla (Valparaíso Region), Rio Maipo wetland (Valparaíso Region).

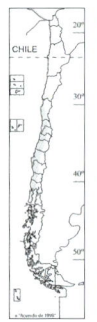

Osprey
Pandion haliaetus   M
L 48cm (♂) 50cm (♀). Sexes similar. **W** 140–190cm. **ID** Unmistakable. When perched, silhouette is very distinctive. Its very long wings exceed the tail, and the bill's shape and hook are different from other raptors. Yellow eyes. **In flight** Long wings give it a rectangular appearance and enhance overall size. The curved wings are characteristic. Appearance is unique among Chilean raptors. **Habitat** Marine bays, lakes or rivers with abundant fish, with high perches in environs. Never visits high-Andean wetlands. **Voice** Largely silent in Chile, but alarm call is occasionally heard, a loud *pewee-pewee-pewee-pewee*. **Where to see** Lluta wetland (Arica y Parinacota Region), Mantagua (Valparaíso Region), Tongoy (Coquimbo Region), Rapel lagoon (O'Higgins Region), Cruces River (Los Ríos Region).

PLATE 51: FALCONS AND KESTREL

cassini

cassini (pale morph)

tundrius

Peregrine Falcon
Falco peregrinus   R
L 42–44cm (♂) 47–50cm (♀). **W** 94–116cm. Sexes similar. **N** 0–4,600m. **C, S & A** 0–3,000m. **ID** Unmistakable. Powerful, deep-chested falcon, with somewhat long, pointed wingtips. Striking deep dark hood, with barred underparts in adult, streaked in juvenile. **In flight** Medium to large-sized falcon. Robust but compact body. Direct fast flight. **Habitat** All manner of habitats from sea level to 4,600m. Coastal cliffs. Cities with tall buildings, especially in winter. **Voice** Mainly silent, except when breeding. All calls basically monosyllabic, varying slightly in pitch and volume, and whether repeated or not. Strident when used to announce territory, from a perch, and repeated 10–12 times (*kack-kack-kack-kack-kack*....) or in courtship, by both pair members in flight. Also in flight, apparently in warning and/or threat, a very sharp and prolonged *keeeeeeeeee*. **Where to see** No specific sites. Morro de Arica in spring/summer (Arica y Parinacota Region), Santiago in winter, Farellones in summer (Metropolitan Region).

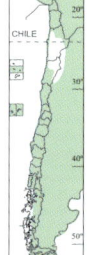

American Kestrel
Falco sparverius   R
L 25–27cm. **W** 52–61cm. Sexually dimorphic. **N** 0–4,000m. **C** 0–2,500m. **S & A** 0–1,200m. **ID** Unmistakable. A relatively small- and slim-bodied falcon, with a relatively long tail and block-shaped head. Sexual dimorphism in plumage very marked; male being especially colourful. **In flight** Small, light-bodied falcon with long pointed wings and long tail. Direct flight on fast wingbeats with short glides. Can hover. **Habitat** Occupies wide variety of habitats, from sea level to 4,000m in altiplano, with marked preference for open areas with prominent perches from which to hunt. **Voice** Silent, except in courtship and when nesting. Like other falcons, both sexes give rapid high-pitched shrieks (*klee-klee-klee-klee-klee-klee-klee-klee-klee*...), the female at lower pitch. **Where to see** Virtually anywhere in range. Common in cities.

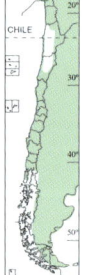

Aplomado Falcon
Falco femoralis   R
L 40cm (♂) 48cm (♀). **W** 78–102cm. Sexes similar. **N** 0–4,600m. **C & S** 0–2,500m. **A** 0–1,500m. **ID** Unmistakable. Proportionately long- and barred-tailed falcon and a white trailing edge to the wing. Striking head pattern immediately identifies the species at rest. **In flight** Medium-sized, light-bodied falcon. Long pointed wings, and long tail. Usually flies fast and direct, unless hunting. **Habitat** Prefers open areas with low vegetation, but occurs in wide range of habitats including small towns. In far south found in Patagonian steppe. **Voice** Generally silent, except when breeding. Mostly heard is a monosyllabic low-intensity call, repeated at slight intervals, during pair interactions or with chicks (*Kip - Kip - Kip - Kip* ...); another, strident and louder, with emphasis on final syllable (*kek-kek-kek-kek-keke-kék*...) appears to be given in aggression or alarm, while another is more sustained, at medium volume (*ki-ki-ki-ki-ki-ki-ki-ki-ki-ki-ki* ...). **Where to see** High Andes in the north: Putre, Socoroma, Chapiquiña, Belen (Arica y Parinacota Region), near La Serena/Ovalle (Coquimbo Region), Pali Ayke National Park, road from Puerto Natales to Torres del Paine National Park (Magallanes Region).

PLATE 52: CARACARAS

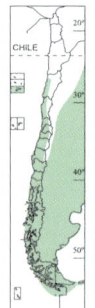

Chimango Caracara
Milvago chimango   R

L 37–40cm. **W** 80–99cm. Sexes similar. 0–2,000m. **ID** Adult similar to juvenile/early immature Mountain and White-throated Caracaras. Pale body and pale grey bill, never black. **In flight** Medium size and inconspicuous coloration. Rather rectangular wings and long tail. Intersperses deep wingbeats and glides. **Habitat** Wide variety of environments. Prefers valleys, agricultural areas and coasts, and is regular in towns and cities. **Voice** Regularly heard, characteristic voice comprises a sustained initial note and descending ones (*cheeeww-cheeww-cheeww-cheew-cheew-cheew*...), given while leaning head backwards, and perched prominently or in flight. **Where to see** Virtually anywhere in range, including cities and near rubbish dumps.

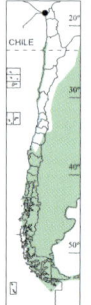

Southern Caracara
Caracara plancus   R

L 55cm (♂) 60cm (♀). **W** 107–133cm. Sexes similar. 0–2,000m. **ID** Largely dark-plumaged raptor, with mainly pale tail and broad white patches in the primaries. Looks capped at rest, with flat crown and large grey bill with bright orange-red base. Juvenile looks more streaked. **In flight** Large. Broad rectangular wings, long white tail with brown terminal band and narrow bars. Slow wingbeats and long glides. **Habitat** Wide variety of environments. In far north only on coast and adjacent small islands. In the south, forests, agricultural land, pampas and coastal regions. **Voice** Regularly heard. In aggression a harsh, dry monosyllabic squawk is given 4–8 times, sometimes ending with bird leaning head backwards while uttering a longer quieter note (*krac-krac -krac-krac-krrrrrrrrrrrrrrr* ...). The name 'caracara', of Guaraní origin, is onomatopoeic. **Where to see** Antofagasta–Tal Tal road (Antofagasta Region), Chungúngo/Punta de Choros (Coquimbo Region), Nahuelbuta National Park (Araucanía Region), Puyehue hot springs (Los Lagos Region), near Punta Arenas and Puerto Natales (Magallanes Region).

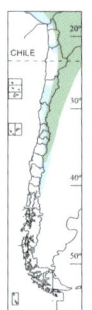

Mountain Caracara
Phalcoboenus megalopterus   R

L 50–57cm. **W** 111–124cm. Sexes similar. **N** 0–4,000m. **C & S** 1,800–3,500m. **ID** Adult unmistakable. Juvenile and immature could be confused with smaller Chimango Caracara. Black bill with grey base and pale blue tone. Also pinkish cere. Has a ragged crest on head. **In flight** Medium-sized. Glides frequently. Long rectangular wings and long tail. In adult black breast contrasts with white underwing and belly. **Habitat** Andes. In far north also seen in lowlands, even on coast (locally). **Voice** Rather silent, but gives high, repeated monosyllabic squawks (*Quiek ... Quiek ... Quiek* ...), sometimes a hoarse moan and a call vaguely reminiscent of Chimango Caracara, but lower and sounding asthmatic. **Where to see** Lauca National Park (Arica y Parinacota Region), road to Talabre (Antofagasta Region), Farellones (Metropolitan Region), road to Lagunillas (Metropolitan Region).

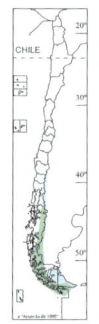

White-throated Caracara
Phalcoboenus albogularis   R

L 50–57cm. **W** 110–119cm. Sexes similar. 0–2,000m. **ID** Adult unmistakable. Juvenile and immature could be confused with smaller Chimango Caracara. Black bill with grey base, and larger and more robust. **In flight** Medium-sized. Smooth wingbeats followed by a glide. Long rectangular wings and long tail. Adult has all-white underparts and inner underwing-coverts. **Habitat** Patagonian forest, meadows and pre-Andean steppe. Perches on cliffs. **Voice** Little information. High-pitched croaks in aggression, usually monosyllabic but sometimes multisyllabic and reminiscent of Chimango and Mountain Caracaras. **Where to see** Torres del Paine National Park and Sierra Baguales (Magallanes Region).

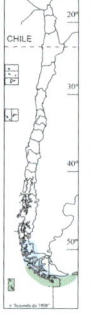

Striated Caracara
Phalcoboenus australis   R

L 59cm (♂) 65cm (♀). **W** 116–125cm. Sexes similar. 0–300m. **ID** Unmistakable. A stocky raptor, with largely dark plumage, except silvery-streaked neck and breast, orange facial skin and rufous thighs and vent. **In flight** White tail tip and patches in primaries, and narrow rufous band on underwing. Medium-sized, but wings broader than other caracaras. Smooth wingbeats followed by a glide. **Habitat** Open areas: islands, beaches, grassland and Patagonian–Andean steppe. **Voice** Varied calls. During aggressive interactions utters monosyllabic, strident harsh squawks (*Khaaa ... Khaaa*), sometimes a rougher, grating *ghaaaaaaaa ... ghaaaaaa* by pair, with head leaning backwards like other caracaras. Pairs also give calls very similar to Southern Caracara, but quieter (*krac-krac-krac*...). **Where to see** Noir Island (Magallanes Region). Accidental on mainland or Navarino Island in winter.

PLATE 53: RAPTORS IN FLIGHT I

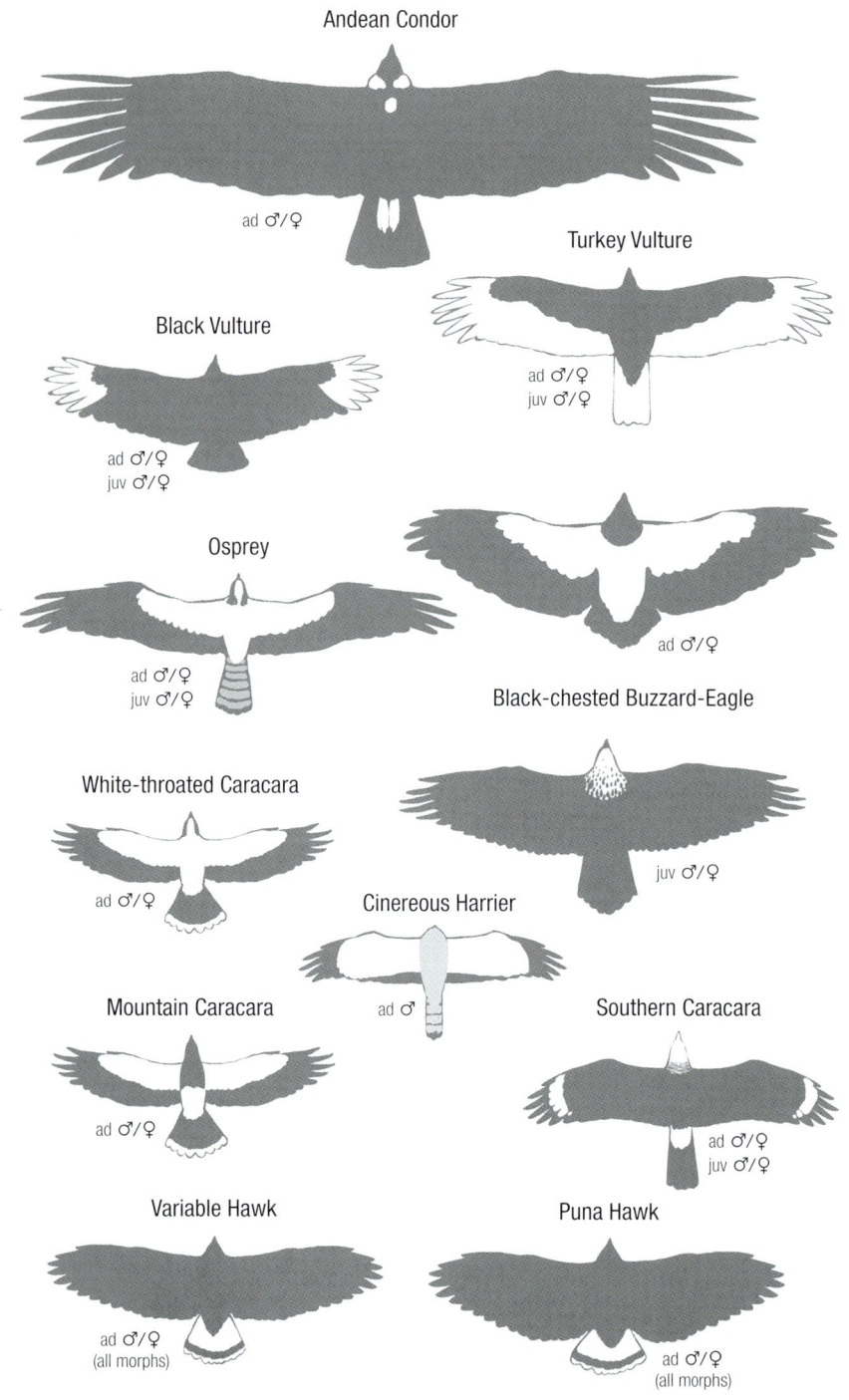

PLATE 54: RAPTORS IN FLIGHT II

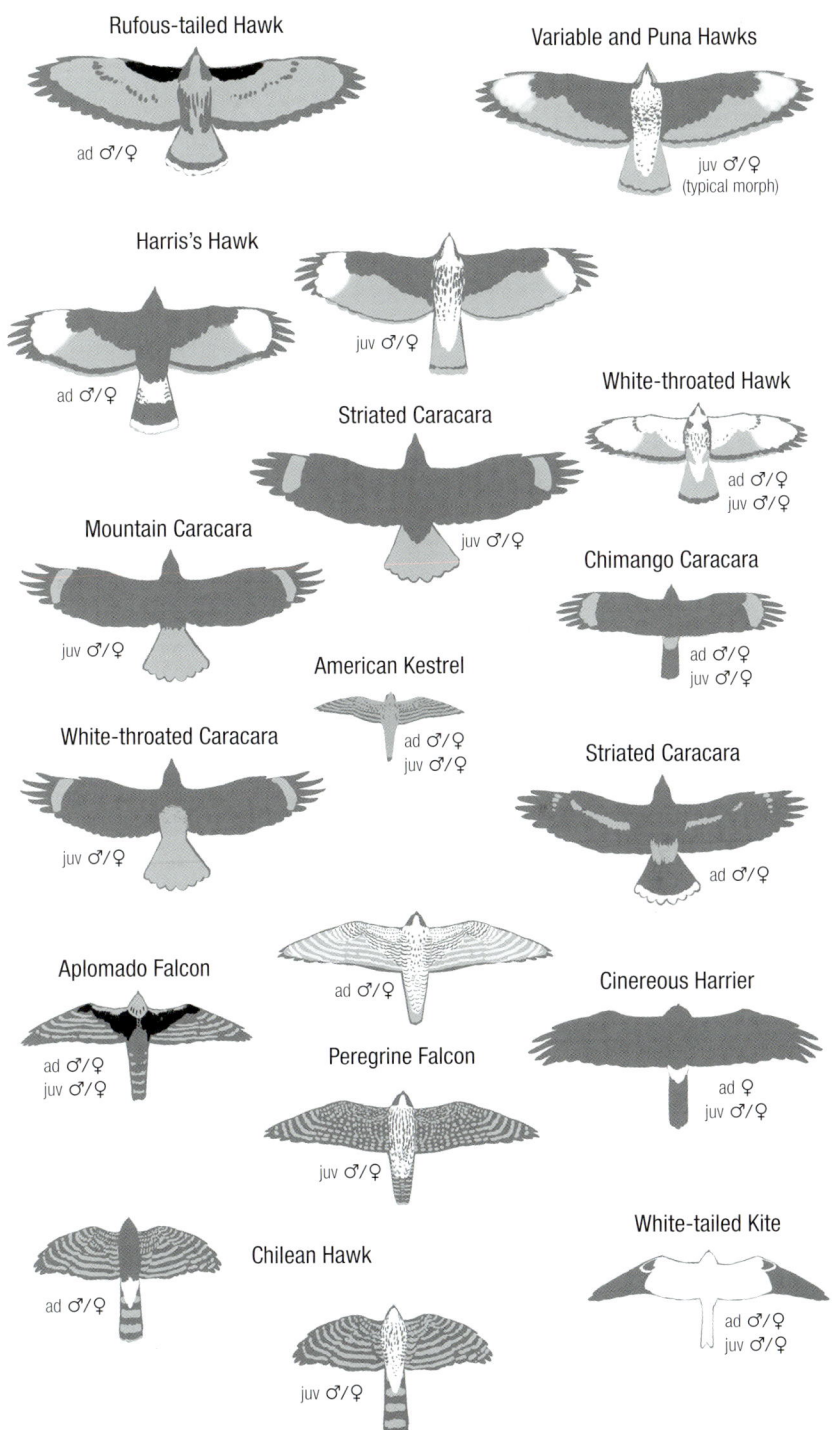

PLATE 55: PIGEONS AND DOVES I

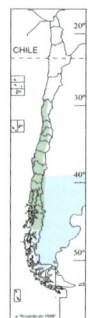

Chilean Pigeon
Patagioenas araucana   **R**
L 35–38cm. Sexes identical. 0–2,200m.
ID Unmistakable perched. In flight could be confused with Rock (Feral) Pigeon, but has no white rump, and uniformly grey plumage. **Habitat** Southern forest, both interior and understorey, and nearby cultivation. In central and northern Chile, also sclerophyll forest, especially in winter. **Voice** Rather silent. A disyllabic, very low-pitched coo given four times (*Hooo… Hoot-hooo Hoot-hooo Hoot-hooo Hoot-hooo…*). Also low grunts in interactions with conspecifics. **Where to see** Zapallar and La Campana National Park (Valparaíso Region), Clarillo River National Reserve (Metropolitan Region), Los Cipreses National Park (O'Higgins Region), Camino Niches (Maule Region), Victoria (Araucanía Region), Chiloé Island (Los Lagos Region).

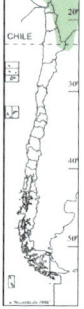

White-winged Pigeon
Patagioenas albipennis   **M/R?**
L 33–34cm. Sexes similar. 0–4,600m.
ID Unmistakable. The only large dove in Chilean highlands. Primary-coverts and alula white, offering obvious contrast in flight. **Habitat** Foothills and mountains in far north. Sometimes in lowlands. **Voice** Low, brief calls likened to growls, moans, or frogs, becoming more aggressive in disputes but always low-pitched. Most frequently heard is a low-frequency, hoarse and raspy series of four syllables, one long note preceding three shorter ones (*ghooooo….. gho - gho - ghooo*), usually given from a high perch. **Where to see** Chapiquiña, Saxamar, Socoroma, Putre (Arica y Parinacota Region). **Note** Nesting very probable, but unproven in Chile. Formerly treated as conspecific with Spot-winged Pigeon *P. maculosa*.

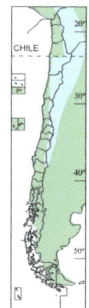

Rock Dove/Feral Pigeon
Columba livia   **Ri**
L 31–33cm. Sexes similar. 0–4,500m. **ID** Similar to Chilean Pigeon in flight, but most individuals have a white rump. **Habitat** Population largely semi-domestic and found in areas with close proximity to humans. **Voice** During courtship males often make cooing. Chicks are very noisy in the nest. **Where to see** Cities and other settlements.

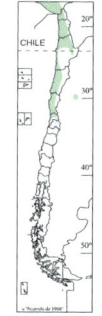

West Peruvian Dove
Zenaida meloda   **R**
L 27–33cm. Sexes similar. 0–2,700m.
ID Unmistakable Noticeable white in wings (secondary-coverts), long tail and blue orbital skin. **Habitat** Arid, semi-arid and irrigated valleys. Also in cities. **Voice** Calls mostly in morning and at dusk, much less during hottest hours; a low, melodic lullaby of 2–3 syllables (*kou-koolii* or *kou-kou-koulii*). **Where to see** Arica, Azapa and Lluta Valleys (Arica y Parinacota Region), Iquique (Tarapacá Region), Antofagasta (Antofagasta Region), Copiapo (Atacama Region), La Serena and Illapel (Coquimbo Region).

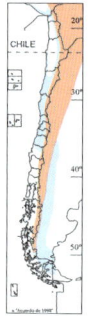

Black-winged Ground-Dove
Metriopelia melanoptera   **R**
L 21–23cm. Sexes similar. **N** 3,000–3,800m. **C** 800–3,000m. **ID** Similar to Eared Dove, but has orange skin around eye and no black on wing-coverts. Obvious white on leading edge of wing in flight. **Habitat** Mountains. Arid and semi-arid country. In far north in scrub below 3,800m. In central Chile below 3,000m. Further south mainly at lower altitudes. **Voice** Most frequently heard call is a frog-like *prrrrweeek-twiu*. Wing noise in flight. **Where to see** Putre, where commoner in winter (Arica y Parinacota Region), Andean Park Juncal (Valparaíso Region), La Campana National Park (Valparaíso Region), Farellones (Metropolitan Region).

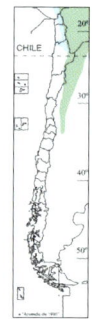

Golden-spotted Ground-Dove
Metriopelia aymara   **R**
L 18–19cm. Sexes similar. 3,500–4,600m.
ID Unmistakable. Plump-bodied, very short-legged and short-tailed dove. Plumage very uniform, with golden spots on lesser coverts and black spots on tertials. In flight, wings and outertail mainly dark. When feeding, looks similar to a rodent, crouched low to the ground. **Habitat** Very arid and stony areas, near waterbodies and steppe. **Voice** Silent in general, with usually all that is heard being the noise of their wings in flight. **Where to see** Parinacota, road to Surire, Parköaylla–Codpa road (Arica y Parinacota Region), road to Paso Jama (Antofagasta Region). In winter, Pachama, Belen, Saxamar, Tignamar (Arica y Parinacota Region).

PLATE 56: DOVES II

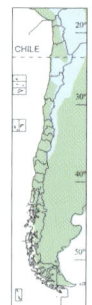

Eared Dove
Zenaida auriculata   R
L 43–46cm. Sexes similar. 0–2,000m. **ID** Similar to Black-winged Ground-Dove. Perched, note blackish spots on wing-coverts and no bare skin around eye. **Habitat** Very adaptable. Arid and semi-arid areas, sclerophyll forest, forested foothills and cities. Shuns dense vegetation, but occurs at woodland edge. **Voice** Low-pitched quiet coos (*khou-khou-khou-khouou*). **Where to see** From Lluta Valley (Arica y Parinacota Region) to southern Chile.

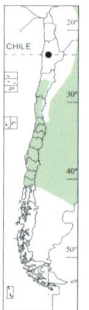

Picui Ground-Dove
Columbina picui   R
L 18–19cm. Sexes similar. 0–2,000m. **ID** Unmistakable. Largely grey with a somewhat long tail. Prominent white in wings and tail, iridescent blue line in wing coverts and very pale underparts. **Habitat** Arid and semi-arid areas with pastures and bushes. Avoids dense vegetation, but does occur in cities. **Voice** A two-note cooing song (*Q'hooop ... Q'hooop...*), which in central Chile is heard even during the hottest hours of summer. **Where to see** Coastal range (Coquimbo Region–O'Higgins Region), Lampa/Batuco system (Metropolitan Region), Rio Maipo wetland (Valparaiso Region), semi-arid parts of central Chile.

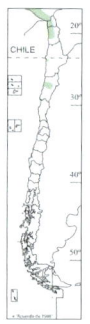

Croaking Ground-Dove
Columbina cruziana   R
L 15–18cm. Sexes similar. 0–1,500m. **ID** Unmistakable. A Ground-Dove with a somewhat long tail and iridescent maroon line on wing coverts. Bill larger than normal for Ground-Doves, especially in males; the basal half is orange. **Habitat** Arid and semi-arid open scrubland. Cities. **Voice** Males gives a raspy call, variably repeating a single note with no real pattern, and recalling a frog (*wrrreeoou*). **Where to see** Lluta and Azapa Valleys (Arica y Parinacota Region), Pica/Matilla (Tarapacá Region).

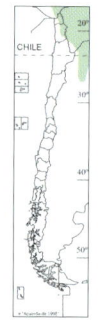

Bare-faced Ground-Dove
Metriopelia ceciliae   R
L 16–18cm. Sexes identical. 2,800–3,800m. **ID** Unmistakable. Small and rounded appearance, with large area of orange-yellow skin around eye. Warm brown body; upperparts and wing-feathers pale-tipped, creating a mottled or scaly pattern. **Habitat** Arid, semi-arid and shrubby areas with little cover. High-Andean villages. **Voice** Most calls rather quiet. Sharp growls in dispute with conspecifics, and a short *wu-wou* in apparent aggression, and a call that vaguely recalls that of Croaking and Black-winged Ground-Doves. On suddenly taking flight, the wings produce a distinctive, metallic sound, like a rattle. **Where to see** Socoroma, Putre, Belen (Arica y Parinacota Region).

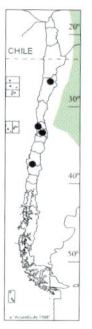

Ruddy Ground-Dove
Columbina talpacoti   V
L 15–18cm. Sexually dimorphic. 0–100m. **ID** Unmistakable. Very small, sexually dimorphic dove. Adult male is largely rufous with grey head. Adult female shares same overall pattern but is obviously duller. **Habitat** Arid and semi-arid areas, scrub and forest edge. Prefers open fields although it can be quite arboreal. **Voice** Mainly heard is a soft, monotonous and low-pitched lullaby (*Hu-Ú, hu-Ú, hu-Ú ...*) very similar to song of Picui Ground-Dove. **Where to see** Accidental in Chile.

PLATE 57: PARROTS

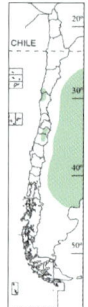

Burrowing Parrot
Cyanoliseus patagonus R
L 43–47cm. Sexes identical. 0–2,000m.
ID Unmistakable. The only Chilean parrot with a largely yellow body. **Habitat** Edges of sclerophyll forest and semi-arid areas, but always near watercourses. **Voice** Noisy. Generally monosyllabic and loud, but often repeated. Most calls dry and short, and very similar. Gives a sharp contact call (*kggghh-kggghh-kggghh...*) while foraging, perched or in steady flight, as well as an alert *Kreeaaa.. Kreeaaa* or *kgaaaaahh-kgaaaaahh-kgaaaaahh-kgaaaaahh....* **Where to see** Santa Gracia Nature Reserve and Montepatria (Coquimbo Region), Los Cipreses National Reserve (O'Higgins Region), near Colbún reservoir (Maule Region).

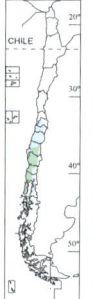

Slender-billed Parakeet
Enicognathus leptorhynchus Re
L 40–42cm. Sexes identical. 0–2,000m. **ID** Similar to Austral Parakeet. Upper mandible obviously long and curved, forehead bright red and extends to the lores. In flight, usually direct without abrupt changes in direction. **Habitat** Southern temperate forest and adjacent open areas. Typically in *Nothofagus* sp. and *Araucaria araucana* forests; occasionally sclerophyll forest in central Chile. **Voice** Very noisy, especially in flocks, either perched or flying. Perhaps most frequently heard is a single monosyllabic *kriiee* or repeated *kriie-kriie-kriie....* Also a repeated *Kraaak ... Kraaak ... Kraaak....* Always clearer and sharper than Austral Parakeet, without any raspy quality. **Where to see** Nahuelbuta range, Ñielol Natural Monument, Conguillio National Park (Araucanía Region), Puyehue National Park (Los Lagos Region).

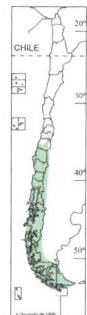

Austral Parakeet
Enicognathus ferrugineus R
L 33–36cm. Sexes identical. 0–2,000m.
ID Compared to Slender-billed Parakeet has shorter upper mandible and a smaller reddish forehead. Flight is much less direct, with frequent rapid changes in direction. **Habitat** Temperate forest. In winter enters sclerophyll forest in central Chile. **Voice** Very noisy, especially in flocks. Most often heard is a monosyllabic call, slightly raspy, shrill and nasal, used in contact both in flight and when perched (*Quiack ... Quiack ... Quiack ...*). **Where to see** Altos de Lircay National Reserve (Maule Region), Termas de Chillán (Ñuble Region), Puyehue National Park, Chiloé Island (Los Lagos Region), Torres del Paine National Park (Arica y Parinacota Region).

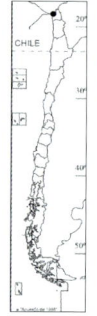

Red-masked Parakeet
Psittacara erythrogenys Ri?
L 32–34cm. Sexes identical. <1,000m.
ID Unmistakable. Large, mainly green parakeet with large white orbital ring and much red on face, and less on edge of wing and on the underwing. **Habitat** Semi-arid areas, sparse woodland and bushes. In Chile, only in valleys around Arica. **Voice** Commonly heard. A high-pitched, powerful squawk, sometimes uttered singly, or repeated 3–4 times (*Kraaaak, Kraaaak, Kraaaak...*), apparently in contact. All calls unmistakably those of parrots, of which this is the only species in Arica. **Where to see** Very local in Azapa Valley, commonly near the Archaeological Museum (Arica y Parinacota Region). **Note** Present since around the year 2000, with a small flock apparently comprising several species, among them *P. erythrogenys*, all presumably escaped captives.

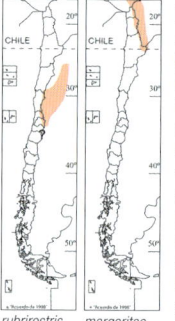

Mountain Parakeet
Psilopsiagon aurifrons R
L 16–19cm. Sexes similar.
N 3,000–4,500m. **C** 2,000–3,500m.
ID Unmistakable. Only parakeet seen above 2,000m altitude in Chile. Has blue in flight feathers, which is not always evident in the field, and small brightly coloured bill in males (grey in females). Green body and noisy character are diagnostic of this species in the environment it inhabits. **Habitat** Mountains and high-Andean environments, near water. **Voice** Calls frequently, to maintain social cohesion within flocks. Their calls draw the attention, and are unmistakably psittacine, which makes identification straightforward as this is the only parrot in its range. Individuals give a *plea-plea-plea ... Plea-plea-plea...*, which is taken up by the flock and acquires a somewhat metallic quality. **Where to see** Socoroma, Putre and Guallatire (Arica y Parinacota Region), near Ollagüe (Antofagasta Region), El Juncal Andean Park (Valparaíso Region), El Yeso Valley (Metropolitan Region).

rubrirostris   margaritae

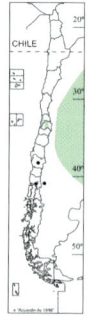

Monk Parakeet
Myiopsitta monachus Ri
L 27–29cm. Sexes identical. 0–1,000m.
ID Unmistakable. Long- and narrow-tailed parakeet, with yellow bill and grey foreface and breast. Very distinctive. **Habitat** Recorded almost exclusively in cities, on towers and metal structures, but also in ornamental Monkey Puzzle (*Araucaria* sp.) trees and ornamental palms, where it builds colonial nests. **Voice** Noisy. A rough, scratchy parrot call given singly (*kraaac ...*) or repeated in alarm (*kraaac-kraaac-kraaac ...*), and may be given by an entire flock, especially in flight to maintain contact. **Where to see** Ñuñoa Square, entrance to PW Country Club, Lo Castro and Lampa (Metropolitan Region).

PLATE 58: OWLS AND BARN OWL

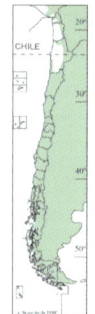

Magellanic Horned Owl
Bubo magellanicus   R
L 43–48cm. Sexes identical. N 0–4,500m. C & S 0–2,500m. A 0–1,500m. ID Unmistakable. The country's largest owl, with very broad wings, a large head but short tail, yellow eyes, and prominent ear-tufts at rest. Facial disc clearly outlined in black. Habitat Open areas with patchy forest. Also rocks and low scrub. Voice Most characteristic is a *hoo-koo-k'rrr*, from which it derives its common name in Chile. Pairs duet in spring, usually with a two-note song, but sometimes three or four (*Hoo-koo - Hoo-koo - Hoo-koo –k'errrr ...*). Apparently in alarm gives *Keck-Keck-Keck-Keck....* Prior to fledging, threatened young clack the bill repeatedly. Where to see Andean foothills (Coquimbo Region–O'Higgins Region), sclerophyll forest (Coquimbo Region–Maule Region), Pali Ayke National Park (Magallanes Region), Farellones–La Parva (Metropolitan Region). Note Although the split from Great Horned Owl *B. virginianus* has been questioned, the obvious difference between vocalisations of birds throughout Chile and southern Argentina compared to those of *B. virginianus* supports their separation.

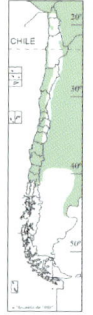

Burrowing Owl
Athene cunicularia   R
L 19–25cm. Sexes identical. N 0–3,200m. C & S 0–1,000m. ID Unmistakable terrestrial owl, with bold head pattern, large yellow eyes, pale-spotted upperparts and mainly barred underparts (largely unmarked in juvenile). Habitat Arid and semi-arid environments, even on sand, provided there is nearby vegetation, even in the Salar de Atacama. In central Chile, a typical bird of very dry environments and coastal sands. Voice Most characteristic is an alarm call, typically when threatened near its burrow and commonly given from a perch c.2m high, before flying off to draw the intruder's attention. This call is high-pitched, and comprises a long initial syllable followed by a short one, repeated variably, sometimes up to eight times (*keeeeek-kee-kee-kee-ke-ke-ke-ke ...*). Another call is given mostly at night and is high-pitched but relatively quiet (*kuh-kuahhhh*). Where to see Lluta and Chaca Valleys (Arica y Parinacota Region), Bosque Fray Jorge National Park (Coquimbo Region), Chacabuco Hills (Valparaíso Region), Lampa/Batuco system (Metropolitan Region), coastal plateaux (Coquimbo Region–Maule Region).

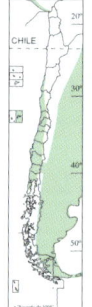

Short-eared Owl
Asio flammeus   R
L 38–42cm. Sexes similar. 0–700m. ID Unmistakable. Medium-large but rather slim-bodied and long-winged owl, with largely pale facial disc and yellow eyes. As its name suggests ear-tufts tiny. Streaked underparts. Most frequently seen in flight, with largely pale underwings. Habitat Plains, swamps and flooded areas, as well as reedbeds around lakes or lagoons. Voice Often silent, but has several vocalisations plus bill-snapping. Most frequently heard is an alarmed *Kieec- Kieec – Kieew*, given both perched and in flight. Where to see Lampa/Batuco system (Metropolitan Region), Chiloé Island (Los Lagos Region), wetlands with reeds.

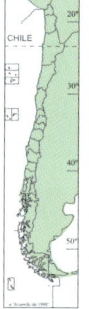

Barn Owl
Tyto alba   R
L 36–40cm. Sexually dimorphic. 0–4,300m. ID Unmistakable. Buff-and-white owl which usually appears virtually all white in flight, and long-legged if seen on ground. Habitat In any environment, except true desert, but generally prefers open areas with patches of old forest. Associated with settlements or abandoned buildings, even inhabiting attics, eaves, etc. Voice Typically heard at night. A long high-pitched *wisshhhhhhhh* given repeatedly; a *Shhhht ... Shhhht ... Shhhht ...*, also repeated; and a metallic patter repeated over and over, both perched and in flight (*tick ... tick ... tick ... tick ... tick ...*). Prior to fledging, young when threatened clack the bill. Where to see Widely distributed, but few regular sites. Note Birds in Chile are sometimes considered to be part of a separate species, American Barn Owl *T. furcata*.

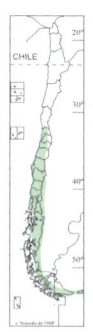

Rufous-legged Owl
Strix rufipes   R
L 33–38cm. Sexes similar. C 0–2,000m. ID Unmistakable. Round-headed and bulky-bodied owl, with strongly barred underparts, dark eyes and a grey facial disc. Habitat Dense mature forests, sclerophyll or temperate. Voice Especially vocal during breeding season. Most frequently heard call, used in territory advertisement, starts low, then becomes higher and louder, sometimes ending in a howl (*ko-ko-ko-ko-KWA-KWA-KWA-KWA-KWA-KWA...*), recalling the screams of a monkey. Other calls resemble a meow and a dry branch cracking, while another involves a repeated syllable (*koh-koh-koh-koh --- koh -koooh...*), sometimes given by both adults, perhaps to reinforce the pair bond. Where to see More often heard than seen. La Campana National Park (Valparaíso Region), Clarillo River (Metropolitan Region), coastal range forests (Valparaíso Region–Maule Region), Termas de Chillán (Ñuble Region), Nahuelbuta National Park and Chiloé Island (Los Lagos Region).

PLATE 59: PYGMY-OWLS, NIGHTJARS AND NIGHTHAWK

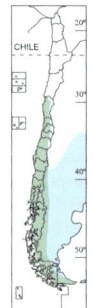

Austral Pygmy-Owl
Glaucidium nana   **R**
L 17–21cm. Sexes identical. 0–2,000m.
ID Unmistakable in range, but Peruvian Pygmy-Owl always separated by vocalisations. **Habitat** Diverse, both temperate and sclerophyll forests, even in cities. Semi-arid areas, but always needs some woodland. **Voice** Frequently heard. Characteristic territorial song comprises a monotonous series of whistles, 2–3 notes per second (*hook-hook-hook-hook- hook-hook-hook...*), at constant volume, commonly heard at night. Calls higher and more melodic, sometimes loud. **Where to see** Anywhere in range, even in large cities.

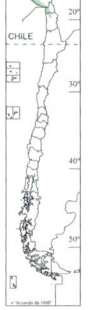

Peruvian Pygmy-Owl
Glaucidium peruanum   **R**
L 15–18cm. Sexes identical. 0–3,600m.
ID Unmistakable in range, but Austral Pygmy-Owl always separated by vocalisations. **Habitat** From semi-desert with sparse vegetation, to cultivation, urban parks and mountainous areas. **Voice** Two main vocalisations. The territorial song is a single note repeated 6–8 times per second (*koi-koi-koi-koi-koi-koi-koi...*). Compared to Austral Pygmy-Owl, it is much faster and more prolonged. Also a higher and louder call, a series of four-syllable phrases that becomes louder and is varied (*kek-kek-kek-kek ... Kll-Kll-Kll-Kll-Kll...*). **Where to see** Codpa Canyon, Lluta, Azapa and Chaca Valleys (Arica y Parinacota Region).

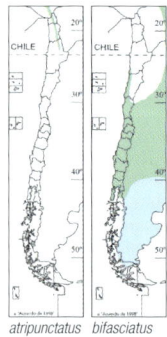

Band-winged Nightjar
Systellura longirostris   **R**
L 21–25cm. Sexually dimorphic. **N** <3,600m. **C** <2,700m. **S & A** <1,200m. **ID** Similar to Lesser Nighthawk and Tschudi's Nightjar. In the north, note rufous hindneck, whitish throat (ssp. *atripunctata*) with long and abundant rictal bristles. Wing and tail markings (white or ochre) relatively large. **Habitat** Steppe, areas of bushy vegetation or woodland with clearings. **Voice** Calls at twilight. Best known call characteristic of austral spring and is sometimes repeated frequently, at

atripunctatus   *bifasciatus*

other times after long intervals, a two-syllable, very high *Cheee-weet ... Cheee-weet....* **Where to see** Road to Socoroma (Arica y Parinacota Region), around Putre (Arica y Parinacota Region), Jere Valley (Antofagasta Region), near Salamanca (Coquimbo Region), Lo Prado (Metropolitan Region), road to Chepu (Los Lagos Region), generally on rural roads at dusk. [Alt. Greater Band-winged Nightjar, Patagonian Nightjar]

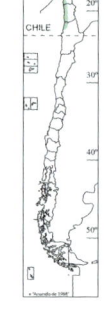

Tschudi's Nightjar
Systellura decussata   **R**
L 20–21cm. Sexually dimorphic. <1,000m.
ID Similar to Band-winged Nightjar. Obviously small with proportionately short wings. Relatively small wing and tail markings (white or ochre). **Habitat** Arid and semi-arid areas, at edge of desert, with scrubby vegetation and Desert Saltgrass (*Distichlis spicata*). **Voice** Calls at twilight. Best known call characteristic of austral spring and is sometimes repeated frequently, at other times after long intervals, a monosyllabic, high *Peeoup ... Peeeoup ... Peeeoup ... Peeeoup....* **Where to see** Lluta and Azapa Valleys, road to Vitor (Arica y Parinacota Region). **Note** Formerly treated as conspecific with Band-winged Nightjar. [Alt. Lesser Band-winged Nightjar]

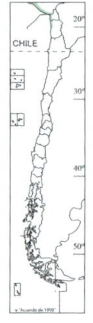

Lesser Nighthawk
Chordeiles acutipennis   **V**
L 19–20cm. Sexually dimorphic. **N** <1,000m.
ID Similar to Band-winged Nightjar. Long-winged with a white (in male) or ochre throat (female), and very short and sparse rictal bristles. Male has narrow white subterminal tail-band. **Habitat** Open country with scrub or wooded clearings. **Voice** Best known is a long invariable trill (lasting up to one minute) that recalls the cooing of a pigeon, or a toad (*uorrrrrrrrrrrrrrr...*). **Where to see** Only accidental in Chile.

PLATE 60: WOODPECKERS, TREERUNNER AND KINGFISHER

Magellanic Woodpecker
Campephilus magellanicus R
L 36–46cm. Sexually dimorphic. 0–2,000m.
ID Unmistakable. Large black-bodied woodpecker, with predominately white underwing, white tips to tertial and inner secondaries and white stripe on upperwing. Noisy and conspicuous, especially the males. Males have entirely red head, while females have a peculiar crest. Both sexes show white remiges on back when perched. Pure white remiges are indicative of adult; if these are black-spotted, it is an immature. **Habitat** Mature forest, already colonised by wood-boring insects. Usually in Southern Beech (*Nothofagus* spp.) and Monkey Puzzle (*Araucaria araucana*) forests. **Voice** Noisy, with drumming and several calls, all strident and loud, always with a nasal quality. Most frequently heard is a repeated, monosyllabic and drawn-out *P'eeaaah ...P'eeaaah...P'eeaaah...* in contact, with another, recalling a human laugh, usually starting with two clearly separated notes, but then becoming a run-together *KEEA-keeaaa-keeaakeeaakeeaakeeaakeeaaa....* Drumming consists of two very loud blows, only easily confused with the noise of an axe. **Where to see** Altos de Lircay National Reserve (Maule Region), Termas de Chillán (Ñuble Region), Nahuelbuta National Park (Arica y Parinacota Region), Torres del Paine National Park (Magallanes Region).

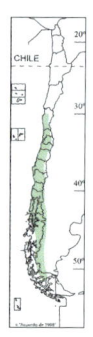

Chilean Flicker
Colaptes pitius R
L 30–33cm. Sexes similar. 0–2,000m.
ID Unmistakable. Very similar to Andean Flicker but allopatric. **Habitat** Open woodland and bushy areas. **Voice** A characteristic sound of Mediterranean Chile. Two well-known calls: the best known comprises two well-separated notes, given either once or repeated several times, the second more accentuated and louder (*Pit-tew ... Pit-tew ... Pit-tew*); the other has a metallic resonance, a single note repeated multiple times, which is given in flight (*klekleklekleklekleklekle....*), vaguely recalling an American Kestrel. Drumming not heard. **Where to see** Road to Farellones and Mahuida Park (Metropolitan Region), La Campana National Park (Valparaíso Region), Los Cipreses National Reserve (O'Higgins Region), Nahuelbuta National Park (Araucanía Region), Chiloé Island (Los Lagos Region).

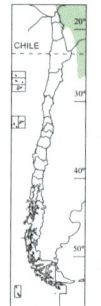

Andean Flicker
Colaptes rupicola R
L 31–34cm. Sexes similar. >4,000m.
ID Unmistakable. Very similar to Chilean Flicker but allopatric. **Habitat** Only in the Andes, near high-Andean bogs (bofedales), rivers and streams, and wet grassland. **Voice** Very noisy and loud, all calls similar in tone. In contact and when alert, gives a monosyllabic and repeated *Kweak... Kweak ... Kweak....*; while in territory advertisement gives a long high *Kikikikikikikikiki...* usually from a perch. Regularly heard in flight is a *Pipipipi ... Pipipipi ... Pipipipi....* **Where to see** Parinacota, Chucuyo, Caquena (Arica y Parinacota Region).

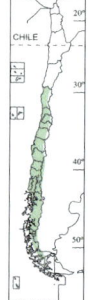

Striped Woodpecker
Veniliornis lignarius R
L 16–18cm. Sexually dimorphic. 0–2,000m.
ID Unmistakable. Smaller woodpecker with heavily streaked underparts, and black-and-white bars above. Striking black-and-white head pattern, augmented in male alone by red nape patch. **Habitat** Relatively open forest, sclerophyll in central Chile or temperate in the south. Also *Acacia caven*, even in parks and woodland in large cities. **Voice** Calls and drumming. Vocalisations simple: in contact, a monosyllabic *Peep... Peep... Peep...* and a complex, loud, high-pitched trill (at 16 notes per second), *kikikikikikikikikirrrrrrrrrr.* Drumming is a very even but very fast tapping at low frequency, by male alone. **Where to see** Road to Farellones and Mahuida Park (Metropolitan Region), La Campana National Park (Valparaíso Region), Los Cipreses National Reserve (O'Higgins Region), Nahuelbuta National Park (Araucanía Region), Pingo River in Torres del Paine National Park (Magallanes Region).

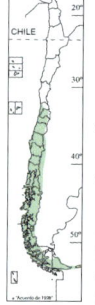

White-throated Treerunner
Pygarrhichas albogularis R
L 15–16cm. Sexes identical. 0–2,100m.
ID Unmistakable. Colour pattern is unique in Chilean forests, as is its behaviour of climbing along trunks, jumping to another tree and always moving. Has rufous/chestnut body and tail with pearl-spotted underparts. Throat and upper breast are pure white. Tail feathers end in rachis (not barbs), like thorns. **Habitat** Forests, mainly southern temperate forest (*Nothofagus* spp.), but also dense sclerophyll forest in central Chile. **Voice** Easily located by its calls. In contact, when moving in pairs, a high somewhat metallic *tick... tick... tick...*, sometimes repeated, or a disyllabic *ticktick... ticktick....* In alarm gives two notes with a slightly metallic quality recalling two stones being tapped together (*Ti-tick... Ti-tick... Ti-tick...*), with many variations. Foraging sounds (tapping trunks and tearing at bark) also quite audible. **Where to see** La Campana National Park (Valparaíso Region), Nahuelbuta National Park (Araucanía Region), Puyehue National Park, Vicente Pérez Rosales National Park (Los Lagos Region), Torres del Paine National Park (Magallanes Region).

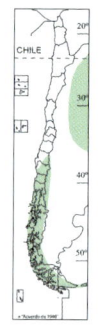

Ringed Kingfisher
Megaceryle torquata R
L 38–44cm. Sexually dimorphic. 0–1,000m.
ID Unmistakable. The only Chilean kingfisher. Its general colour pattern (mainly blue and red) and shape are unique. Has an irregular disordered crest, from bill to nape, and barred tail. **Habitat** Waterbodies, rivers, streams, bays and marine channels. **Voice** Noisy. It emits several calls, relatively similar. Pairs sometimes duet. A persistently repeated note, recalling a rattle or two pieces of wood being hit together (*kek-kek-kek-kek-kek...*), a high trill (*crrrriiiiiii ....*), which can become a somewhat dry, metallic rattle, given for long periods (*kakakakakakaka...*). **Where to see** Chiloé Island (Los Lagos Region), coastal fjords (Aysén Region).

PLATE 61: SWALLOWS

Small birds, highly gregarious and some are long-distance migrants. These characteristics force the need to review all the flocks that fly over feeding, since in the middle of them there may be individuals of an unusual species. This is more relevant in the north of the country.

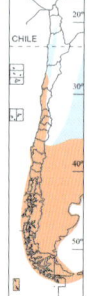

Chilean Swallow
Tachycineta leucopyga　　　　　　　　R

L 12–14cm. Sexes identical. 0–1,500m. **ID** Similar to Blue-and-white Swallow, but has white rump and undertail. **Habitat** Open areas and wetlands, including cities and regularly nests on artificial constructions. **Voice** Frequently heard. Gives a slightly metallic, rather low and somewhat nasal *trwee-teet* perched or in flight, apparently in contact, and sometimes becomes a raspy, hoarse *trrreeee*. Also a very characteristic twitter lasting c.2 seconds, usually when perched and perhaps in aggressive interactions, which comprises c.9 syllables (*twe-teet-teeteetrrriiiii- teetchi-tut* etc.). **Where to see** Fairly common. Lampa/Batuco system (Metropolitan Region), Cartagena lagoon (Valparaíso Region).

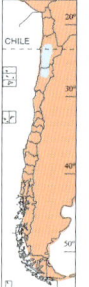

Blue-and-white Swallow
Pygochelidon cyanoleuca　　　　　　R

L 12–13cm. Sexes identical. **N** 0–4,000m. **C** 0–3,000m. **S & A** 0–1,500m. **ID** Similar to Chilean and Andean Swallows, but rump and undertail are both black. **Habitat** Open areas throughout Chile, from mountain valleys to beaches in the south. Often near settlements. **Voice** Quiet but frequent calls, including a trill, from a perch or in flight (*trrrreeep*), sometimes mixed with a chirping sound (*teeeewww*), as well as a variety of other calls, some complex, and in interactions with conspecifics. **Where to see** Lluta and Azapa Valleys (Arica y Parinacota Region), central coastal plateaus (Coquimbo Region–O'Higgins Region), Lampa/Batuco system, Farellones and El Yeso Valley (Metropolitan Region).

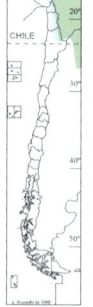

Andean Swallow
Orochelidon andecola　　　　　　　　R

L 14–15cm. Sexes identical. 2,700–4,600m. **ID** Similar to Blue-and-white Swallow, but has smoky-brown plumage, which makes entire head look dark. Juvenile can be also confused with Blue-and-white Swallow, but note pale, whitish rump even at a distance. **Habitat** Open highlands, often around watercourses, lagoons or pools, and high-Andean bogs (bofedales). Perches and nests on cliffs and banks, never very high, but always difficult to access. **Voice** Quiet but frequent calls. Perched flocks give a diversity of soft calls, audible only when close. In flight short twitter-like *trrrwee* and *trreep* calls, which are sometimes mixed. **Where to see** Putre Las Cuevas road, Parinacota, Chungará lake to Tambo Quemado and Guallatire (Arica y Parinacota Region).

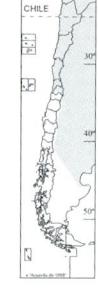

Bank Swallow/Sand Martin
Riparia riparia　　　　　　　　　　　M

L 12–13cm. Sexes identical. **N** 0–4,500m. **ID** Upperparts and breast-band brown. **Habitat** Open areas; not in cities or forested areas. Often over small wetlands. **Voice** In flight gives a short, somewhat hoarse and quiet *tchktchktchktck* ..., quite different from other swallows. **Where to see** No regular sites, but mostly seen in the north (Arica y Parinacota Region–Antofagasta Region). [Alt. Collared Sand Martin]

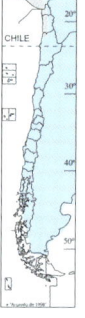

Barn Swallow
Hirundo rustica　　　　　　　　　　M

L 15–17cm. Sexually dimorphic. 0–4,000m. **ID** Similar to Blue-and-white, Cliff and Andean Swallows. A long- and fork-tailed swallow, with white central band. Throat and body dirty ochre to rufous. **Habitat** Open country, from around plantations to high Andes. **Voice** Quiet but frequent calls. Short notes, probably in contact, given in flight or sometimes when perched and interacting with conspecifics. In flight, frequently gives a metallic *plea-plea ... Plea-plea...*, and when perched sometimes utters a whistle-like *tew-weet... tew-weet....* **Where to see** Lluta Valley (Arica y Parinacota Region), Pica/Matilla (Tarapacá Region).

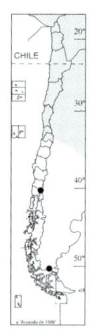

Cliff Swallow
Petrochelidon pyrrhonota　　　　　　M

L 13–14cm. Sexes identical. 0–3,700m. **ID** Similar to Blue-and-white, Barn and Andean Swallows. Compact, with a streaked blue-brown back, ochre rump and hindcollar, and reddish or rufous throat and neck-sides. **Habitat** Open areas and livestock farms. **Voice** Most frequently heard in Chile is a somewhat high-pitched call, *tree...* or *treee-tree*, given in flight, and somewhat reminiscent of a Budgerigar *Melopsittacus undulatus*. **Where to see** Lluta valley (Arica y Parinacota Region).

PLATE 62: MARTIN, SWIFTS AND HUMMINGBIRDS I

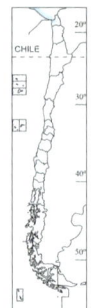

Peruvian Martin
Progne murphyi **V/M?**
L 16–17cm. Sexually dimorphic. 0–1,000m. **ID** Both sexes unmistakable. Adult male is entirely dark blue with duskier underparts and a well-notched tail. Adult female is also all dark, with dusky collar, and shows some blue iridescence on back and scapulars. **Habitat** Open areas, near coasts. **Voice** No information. **Where to see** Few records. Chacalluta airport, Lluta Valley (Arica y Parinacota Region).

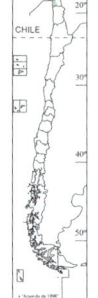

Chilean Woodstar
Eulidia yarrellii **R**
L 7–8cm. Sexually dimorphic. 0–2,600m. **ID** Male unmistakable. Female very similar to female Peruvian Sheartail. Whitish belly, upperparts with golden fringes and tail with inconspicuous cinnamon base. Only tips of rectrices are white. **Habitat** Gorges with native vegetation. **Voice** Produces sounds with wings and tail. Male gives a very short and simple, high-pitched monosyllabic *trrrrrrrrrrrrrt*, either perched or, more commonly, in flight. **Where to see** Increasingly rare. Azapa and Chaca Valleys (Arica y Parinacota Region).

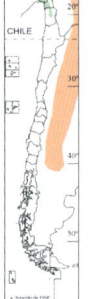

Andean Swift
Aeronautes andecolus **M/R?**
L 13–15cm. Sexes identical. 0–4,500m. **ID** Unmistakable. Long thin wings, and very fast flight. **Habitat** Mostly in foothills and semi-arid mountains, over agricultural areas, Queñoa forests (*Polylepis rugulosa*), etc. **Voice** High-pitched calls, given frequently, often serve to alert observers to its presence, especially a slightly hissing sound that recalls an insect (*tziitziitziitziitziitzii* ...) or a more trilled *triitriitriitriitriitrii*. **Where to see** Molinos, Pachama and pre-puna zone (Arica y Parinacota Region). **Note** Nesting extremely likely, but currently unproven in Chile.

Oasis Hummingbird
Rhodopis vesper **R**
L 11–14cm. Sexually dimorphic. 0–3,500m. **ID** Unmistakable. Both sexes have orange-ochre rump, and long, slightly decurved bill. **Habitat** Gorges with bushes, oases and coastal scrub. Also in cities. **Voice** Infrequently heard. A dry call can be given singly (*tick*) or as a brief, variable series (e.g. *tick...... tick... tick....* or *tick... tick.. tick.. tick...*). More metallic and harder-sounding than Peruvian Sheartail. Also an apparently aggressive, and quickly repeated, very fast trill (*tritritri ... tritritri ... tritritri ...*). In courtship display, male produces sounds with the wings and tail. **Where to see** Azapa Valley (Arica y Parinacota Region), Pica, Matilla and Iquique (Tarapacá Region), Caldera (Atacama Region), La Serena and environs (Coquimbo Region).

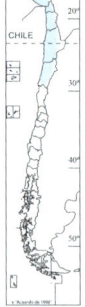

Chimney Swift
Chaetura pelagica **V**
L 12–14cm. Sexes identical. 0–2,700m. **ID** Might recall a swallow or martin in flight, but has an all-dark body and wings. **Habitat** Varied, but seems to prefer areas with watercourses. **Voice** High-pitched trills (*tritri-tri-tri-tri-tri-triiii* ...) in which individual notes are distinguishable; somewhat reminiscent of Oasis Hummingbird. **Where to see** Only in far north: Arica (Arica y Parinacota Region), María Elena, San Pedro de Atacama (Antofagasta Region).

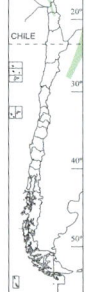

Sparkling Violetear
Colibri coruscans **R**
L 13–14cm. Sexes identical. 2,500–3,600m. **ID** Unmistakable. The only Chilean hummingbird with solidly dark underparts. Relatively large and strong-bodied, principally brilliant green, with a mid-length bill, iridescent blue mask and belly patch, and a grey belly. **Habitat** Open scrub of Yara (*Dunalia spinosa*) and especially *Eucalyptus* sp. **Voice** Calls frequently. Sings from a high perch, a monotonous but characteristic, high-pitched, monosyllabic *tchin, tchin, tchin, tchin*, repeated almost endlessly. Sometimes a lower *trrrt, trrrt, trrrt, trrrt, trrrt, trrrt ...*, also from a high perch. Calls become more complex in agonistic interactions, and more similar to those of other hummingbirds. **Where to see** Socoroma, Putre, Chapiquiña and Ticnamar, all in pre-puna zona (Arica y Parinacota Region).

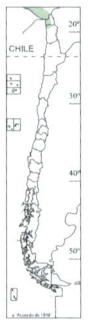

Peruvian Sheartail
Thaumastura cora **R**
L 13–17cm (♂) 7–9cm (♀). Sexually dimorphic. 0–3,000m. **ID** Male unmistakable. Female very similar to female Chilean Woodstar, but has green central rectrices, and the two immediately adjacent feathers either side are white. **Habitat** Gorges with scrub, oases, coastal scrub and agricultural fields. Also suburban areas. **Voice** Calls frequently; unexpectedly loud. Mostly heard is a dry note repeated frequently in different combinations (*tchip-tchip ...... tchip-tchip-tchip....*). Also high trills, which in disputes with other hummingbirds it may introduce some *tchip* notes. **Where to see** San Miguel de Azapa Archaeological Museum (Arica y Parinacota Region).

PLATE 63: HUMMINGBIRDS II

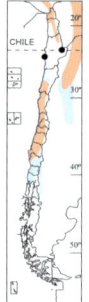

Giant Hummingbird
Patagona gigas R
L 20–23cm. Sexes identical. **N** 2,000–3,800m. **C & S** 0–2,500m. **ID** Unmistakable. Largest hummingbird in the world, with a long bill and long, well-notched tail. Upperparts dull olive-brown with somewhat irregular but obvious white rump, and largely dark brown underparts. All of its movements are typical of a hummingbird, but noticeably slower. Shape, size and colour are diagnostic of this species. **Habitat** Semi-arid areas, Andean foothills and pre-puna (north) and open areas with scrub and very dry vegetation (central Chile). **Voice** Short, shrill monosyllables, singly or repeatedly, but always at intervals (*tzeek …… tzeek …… tzeek …*). Also low, harsh notes. **Where to see** Socoroma and Putre (Arica y Parinacota Region), plateaux and coastal slopes (Coquimbo Region–O'Higgins Region), road to Caleu (Metropolitan Region). Coastal mountains with *Puya* spp. or *Lobelia* spp.

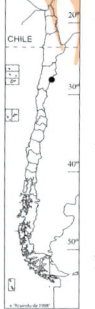

Andean Hillstar
Oreotrochilus estella R
L 12–14cm. Sexually dimorphic. 2,000–3,500m. **ID** Unmistakable by range. Very similar to allopatric White-sided Hillstar. Male has reddish-brown ventral line. **Habitat** Pre-puna, crop fields and Queñoa forest (*Polylepis rugulosa*). **Voice** Not very vocal. A high-pitched *tsik… tsik…* is commonly given once but sometimes repeated 3–4 times, and a high, repetitive *pew-pew-pew-pew-pew…* with variations in tone. **Where to see** Socoroma and Putre (Arica y Parinacota Region), Chusmiza (Tarapacá Region).

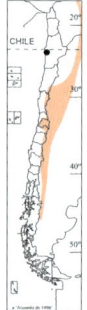

White-sided Hillstar
Oreotrochilus leucopleurus R
L 11–15cm. Sexually dimorphic. 2,000–3,500m. **ID** Unmistakable by range. Very similar to allopatric Andean Hillstar. Male has dark blue ventral line. **Habitat** Andes and coastal range. Usually near water. **Voice** Not very vocal. A series of high-pitched notes, similar to Andean Hillstar. **Where to see** El Juncal Andean Park (Valparaíso Region), Farellones and Maipo canyon (Metropolitan Region).

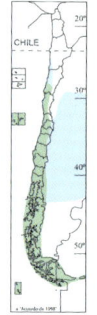

Green-backed Firecrown
Sephanoides sephaniodes R
L 10–11cm. Sexually dimorphic. 0–2,000m. **ID** Unmistakable. Southernmost hummingbird in the world. Largely green hummingbird; both sexes have somewhat streaked throat, speckled green underparts and a small white postocular spot. Male has a brilliant red forecrown patch. Typically the only hummingbird 40°S to Cape Horn. **Habitat** From southern temperate forest to very dry environments in north of range (mostly winter). **Voice** Almost constantly gives high, strident and metallic trills, or twittering notes, singly or in sequences, which can be inaudible to some people. **Where to see** Almost anywhere between the Araucanía Region and Aysén Region in summer. In winter, common in central Chile, mainly on coast (Coquimbo Region–O'Higgins Region), associated with *Eucalyptus* sp. and native Fuchsia (*Fuchsia lycioides*). Also Juan Fernández Islands (Valparaíso Region).

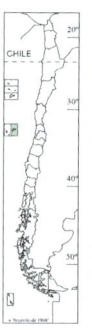

Juan Fernandez Firecrown
Sephanoides fernandensis Re
L 11–13cm. Sexually dimorphic. 0–100m. **ID** Unmistakable. Adult male is stunning, virtually all-rufous with ruby-red forehead, whereas female is mainly deep green above, with largely clean white underparts and brilliant blue forecrown. **Habitat** Native forest and ornamental shrubs and flowers. **Voice** Varied calls, including high, metallic trills similar to Green-backed Firecrown, a high twitter and longer trills. **Where to see** Robinson Crusoe Island, in Juan Fernández archipelago (Valparaíso Region).

PLATE 64: TAPACULOS I

Mostly terrestrial and always more often heard than seen. Vocalisations often loud and very characteristic, making them important for both identification and location.

Chestnut-throated Huet-huet
Pteroptochos castaneus **R**
L 24–25cm. Sexes identical. **C & S** 0–1,800m. **ID** Unmistakable. White eye-ring, and reddish-brown throat, foreneck and breast. **Habitat** Wet and dark forest floor. **Voice** Unmistakable within species' range. In mild alarm, utters a monosyllabic and loud, repeated *Wet!... Wet!...* every five or more seconds, which in true alarm becomes disyllabic, higher-pitched and is repeated several times, usually 4–6, in quick succession, *Wet-wet...Wet-wet...Wet-wet... Wet-wet....* In territory advertisement gives a long series of low *hoot* notes, repeated c.25 times in crescendo, of which first 3–5 are clearly separated, before accelerating, *hoot-hoot-hoot- hoot - hoot-hoot-hoot hoot-hoot....* **Where to see** Altos del Lircay National Reserve (Maule Region), Termas de Chillán (Ñuble Region). **Note** Near-endemic; a small population occurs in Argentina, around Epulafquen lagoons, very near the Chile border.

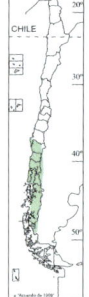

Black-throated Huet-huet
Pteroptochos tarnii **R**
L 22–25cm. Sexes identical. **S & A** 0–1,200m. **ID** Unmistakable. White eye-ring, with blackish-grey throat, foreneck and breast. **Habitat** Wet and dark forest floor. **Voice** Loud and unmistakable in range. In alarm a high-pitched phrase, repeated several times, *wet-wet... wet-wet-wet... wet....* In territory advertisement a single note repeated c.20 times, the first 3–4 iterations longer and well-separated, before accelerating to a plateau (*huut — huut — huut - hut--hut-hut-hut hut-hut...*). Another apparently territorial call involves 15–17 notes that decrease in volume (*WOK-WOK....Wok-Wok-Wok -Wok-Wok- Wok-Wok...*) and vaguely recalls the song of Moustached Turca. Also a low-tone, scratchy *kouuu-kouuu-kouuu-kouuu-kouuu....* Male calls apparently lower-pitched than female. **Where to see** Nahuelbuta National Park (Araucanía Region), Conguillio National Park (Araucanía Region), Alerce Andino National Park, Chiloé National Park (Los Lagos Region).

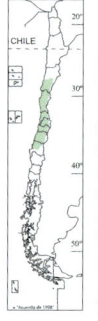

Moustached Turca
Pteroptochos megapodius **Re**
L 23–24cm. Sexes identical. **N** 0–3,800m. **C** 0–2,400m. **ID** Unmistakable. White cheek patches. **Habitat** Common in very dry habitats, montane slopes with scattered vegetation. Avoids dense sclerophyllous forest. **Voice** Two vocalisations are particularly characteristic and commonly heard. One, a territorial call, comprises 10–12 whistles, at the rate of almost one per second, gradually decreasing in volume and pitch, and sometimes lengthening (*Wuut..Wuut-wuut-wuut -wuut- ... -wuuout-wuuout ...*), and the whole sometimes becoming faster. The other, also apparently territorial, is a series of metallic notes, repeated c.20 times (*wok-wokwok-wok-wok-wok-wok...*). **Where to see** La Campana National Park (Valparaíso Region), Farellones road, El Yeso Reservoir road (Metropolitan Region).

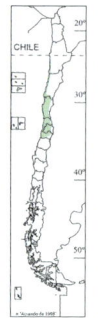

White-throated Tapaculo
Scelorchilus albicollis **Re**
L 18–20cm. Sexes identical. **C** 0–1,700m. **ID** Unmistakable. White throat and breast, plus barred belly. **Habitat** Semi-arid gorges and rocky hillsides. Xerophytic scrub, never far from water. **Voice** Song unmistakable and comprises a three-note phrase with a strongly accented first syllable (*Tá-pa-ku... Tá-pa-ku... Tá-pa-ku...*), typically given from atop a rock, or open branch. More usually given from within cover, starting quietly, and descending further towards the end is a *kow-kow-kow-kow-koww...* or a much clearer and higher-pitched *Kóow... Koow...* that is vaguely cat-like. Two calls are regularly heard, one like a rubber toy being squeezed (*Queeeck*) and the other recalling some pigeons (*bgurrrrrrrrrrr... bgurrrrrrr...*). **Where to see** Santa Gracia (Coquimbo Region), Córdoba gorge (Valparaíso Region), Mahuida Park (Metropolitan Region).

PLATE 65: TAPACULOS II

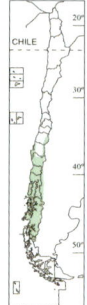

Chucao Tapaculo
Scelorchilus rubecula　　　　　　　　　R
L 17–19cm. Sexes identical. **S** 0–2,000m. **ID** Unmistakable. Reddish breast and black-and-white bars on belly. **Habitat** Floor of *Nothofagus* forest including more open woodland, but prefers dense, wet and cold forest with a dense understorey. **Voice** Wide repertoire. In territory defence gives a loud series of 6–7 notes, the first rather quiet, the next 3–4 very strong and fast, and the last two descending (*Crrr-Pe-Pe-Pe-Pe-ka-kow*...). Also a variable groaning *Kwoow... Kwoow...*, with a whispered quality. **Where to see** Nahuelbuta National Park, Conguillio National Park (Araucanía Region), Termas de Puyehue, Chiloé National Park (Los Lagos Region).

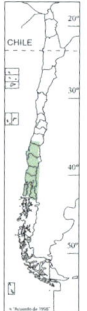

Ochre-flanked Tapaculo
Eugralla paradoxa　　　　　　　　　R
L 14–16cm. Sexes identical. **S** 0–1,000m. **ID** Compare with Magellanic Tapaculo. Flanks, vent and undertail-coverts ochre, with yellowish legs. **Habitat** Wet areas with dense vegetation, often near watercourses. **Voice** Most vocalisations are rather similar, and perhaps merely variations. A repeated note given 5–8 times, forming a long phrase that sounds run-together, strident, dry and metallic (*ChewChewChewChewChewChew*...), or a similar, but more clearly separated, and slower *Chiuk-Chiuk-Chiuk-Chiuk-Chiuk-Chiuk*..., which is equally strident but even drier and more metallic, also repeated c.8 times. Both calls are audible over some distance and recall two stones being hit together. **Where to see** Termas de Chillán (Ñuble Region), Nahuelbuta National Park, Conguillio National Park (Araucanía Region), Puyehue National Park, Chiloé Island (Los Lagos Region).

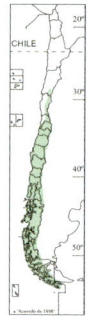

Magellanic Tapaculo
Scytalopus magellanicus　　　　　　　　　R
L 11–13cm. Sexes identical. **C** 2,000–3,500m. **S & A** 0–1,500m. **ID** Could be confused with Dusky Tapaculo in central Chile and Ochre-flanked Tapaculo in the south. Voice is very useful. In the south (typical morph), has slate-grey body, with a variable white spot on the head. Andean morph (over 2,000m; central Chile), has blackish to brownish wings and no white spot on the head. **Habitat** Scrub and wet areas around water (central Chile). In the south, dense vegetation. **Voice** In alarm a sharp and aggressive trill (*treetreetreeop... treetreetreeop... treetreetreeop*...). Song of lowland and southern birds is a highly characteristic and long, repeated *P'trae ... P'trae... P'trae.....* **Where to see** High parts of Maipo canyon (Metropolitan Region), Nahuelbuta National Park (Araucanía Region), Chiloé Island (Los Lagos Region).

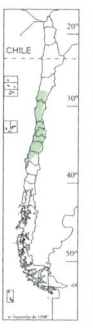

Dusky Tapaculo
Scytalopus fuscus　　　　　　　　　Re
L 11–12cm. Sexes identical. **C** 0–2,600m. **ID** Similar to Magellanic Tapaculo; voice is the best character for identification. **Habitat** Humid and shady ravines and gorges, with abundant bushy vegetation, and sectors of low dense vegetation on banks and river drains. **Voice** Most characteristic is a series of notes, at rate of one every c.0.8 seconds, typically repeated up to 40 times but occasionally as many as 100, always at a steady pace (*T'chorreen... T'chorreen ... T'chorreen ....*). Also, when alert, a high-pitched but quiet and somewhat scratchy, well-spaced *Treuup... Treuup...*), as well as a hoarse *Crrrkkk... Crrrkkk... Crrrkkk... Crrrkkk...* in full alarm. **Where to see** La Campana National Park, Rio Maipo wetland (Valparaíso Region), road to Farellones (Metropolitan Region).

PLATE 66: CINCLODES

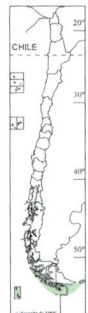

Black Cinclodes
Cinclodes maculirostris   **R**
L 18–23cm. Sexes identical. **A** 0–200m. **ID** Unmistakable. No white in plumage. **Habitat** Coasts, especially with breeding colonies of birds or mammals. **Voice** Short monosyllables, sharp but scratchy in contact voices (*trrweet* ...), and a continuous high-pitched, drawn-out whistle (*Fiifiiiifiiifiiifiiii... Fiifiiiifiiifiiifiiii...*), somewhat reminiscent of Plain-mantled Tit-Spinetail. Song complex. **Where to see** Noir Island, Cape Horn and Diego Ramírez Islands (Magallanes Region). **Note** Formerly treated as conspecific with Blackish Cinclodes *C. antarcticus* of the Falklands. [Alt. Fuegian Cinclodes]

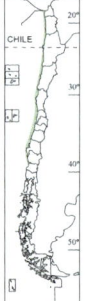

Seaside Cinclodes
Cinclodes nigrofumosus   **Re**
L 22–27cm. Sexes identical. **C** 0–50m. **ID** Similar to Dark-bellied Cinclodes. Very dark with a very narrow ochre superciliary and a large white throat. **Habitat** Exclusively on coasts. Rocky areas, breakers and islets. **Voice** Song starts with a very high note, followed by a strident sequence of the same syllable repeated c.15 times, which gradually decreases in intensity after the sixth repetition (*TWEET-TWEET-Tweet-Tweet-Tweet-Tweet-Tweet-tweet-tweet-tweet...*). **Where to see** Punta Patache (Tarapacá Region), Cachagua and Viña del Mar–Concón road (Valparaíso Region).

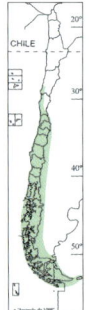

Dark-bellied Cinclodes
Cinclodes patagonicus   **R**
L 19–22cm. Sexes identical. **C** 0–2,600m. **S & A** 0–1,000m. **ID** Similar to Grey-flanked and Seaside Cinclodes. Overall dark, with white throat and superciliary, both reaching well back on head. Flanks concolorous with breast. **Habitat** Coasts, around estuaries and streams, even in cities. **Voice** Similar to other cinclodes. Song has two parts; the first consists of 3–5 monosyllabic notes (the first slightly different) followed by a long sharp trill that finally descends (*Tweee-Twee-Twee-Twee-tee-trrrrrrrrrrrrrrrrrrrrrr* ...). **Where to see** Budi lake (Bio-Bio Region), Chacao and Caulín (Los Lagos Region), Fuerte Bulnes road (Magallanes Region).

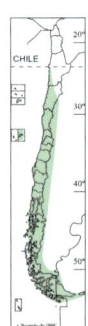

Grey-flanked Cinclodes
Cinclodes oustaleti   **R**
L 16–18cm. Sexes identical. **C** 0–4,200m. **ID** Similar to Dark-bellied and Buff-winged Cinclodes. Dark upperparts (somewhat variable) and whitish belly. Small sharp bill. Flanks chocolate-brown, warm and contrasting with duller brown breast. **Habitat** Wetlands and Andean grasslands in central Chile, and to lesser extent coastal areas, especially in winter. In the south it is more clearly tied to coastal areas. **Voice** Song is a high trill rising slightly at end, preceded by 2–3 distinct notes, *Pip ... Pip ... Pip ... Trweeeeeeeeeeeeeeeet....* **Where to see** Farellones, El Yeso Reservoir (Metropolitan Region). In winter, at coastal wetlands. On Alexander Selkirk Island (Juan Fernández Islands, Valparaíso Region) ssp. *baeckstroemii*.

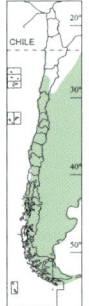

Buff-winged Cinclodes
Cinclodes fuscus   **R**
L 16–18cm. Sexes identical. **C** 0–3,000m. **A** <1,000m. **ID** Similar to Grey-flanked and Cream-winged Cinclodes. Ochre-brown upperparts. White throat grades into scales on breast, sometimes with short whitish streaks on belly. **Habitat** Grassland and areas near wetlands, mostly in Andes. More abundant in lowlands in winter. In the south, more tied to coastal areas. **Voice** Several calls. Song varies, but is usually a sharp, stable trill lasting c.10 seconds, preceded by 3–4 separate notes. **Where to see** El Yeso Reservoir, Farellones, Lampa/Batuco system (Metropolitan Region).

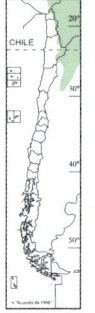

Cream-winged Cinclodes
Cinclodes albiventris   **R**
L 16–18cm. Sexes identical. **N** 2,200–4,600m. **ID** Similar to Buff-winged and White-winged Cinclodes. Overall brown, somewhat dark, with white to creamy-white wing markings, and creamy-white tips to outer tail feathers. **Habitat** Hillsides and high plateaux, around all types of wetlands, especially high-Andean bogs (bofedales), lakes and streams. **Voice** Song has three well-defined parts, starts with a few single notes, then accelerates into a high-pitched trill, before returning to more melodic single notes, which may be repeated many times (*Tuit..tuit... tweet... tweet-tuit-tuit-trrrrrrrrrrrrriiiiiiiiiii-Tuitui-tui-tui-tuitui-trrrrrrrrriiiiiii...*). **Where to see** Putre village, Parinacota and Guallatire bofedales (Arica y Parinacota Region); Machuca bofedales (Antofagasta Region).

White-winged Cinclodes
Cinclodes atacamensis   **R**
L 21–23cm. Sexes identical. **N** 2,500–4,600m. **ID** Similar to Cream-winged Cinclodes. Somewhat larger, reddish-brown, with white wing-bands and pure white tips to outer tail feathers. **Habitat** Always around high-Andean wetlands, especially rivers, streams and high-Andean bogs (bofedales). **Voice** Contact call is a somewhat sharp, short and dry note that is repeated (*Tuiitit ... Tuiitit ...*), and the same note is used to open the song, which then lengthens and accelerates, and is repeated c.4 times (*Tuiitit... Tuiitit... Tuiiiititititit-tuiiiitititit-tuiiiititititit-tuiiiititititit....*), sometimes with the addition of a third part that is almost a trill (*Tuiitit... Tuiitit.. Tuiititititrrrrrrrrrr...*). **Where to see** Parinacota and Chungará lake (Arica y Parinacota Region), Salar del Huasco (Tarapacá Region).

PLATE 67: MINERS

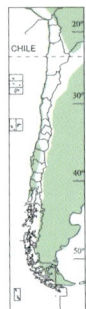

Common Miner
Geositta cunicularia R
L 15–16cm. Sexes identical. **N** 4,000–4,600m. **C, S & A** 0–2,000m. **ID** Similar to other miners. Upper breast streaked brown. Ear-coverts dark brown. Long bill, slightly curved. At rest wings shorter than tail, as primaries very short. In flight, orange band on wing. Tail has pale base (sometimes with pale orange tone) and rest is blackish, forming a triangle. **Habitat** Open country with low vegetation, stony areas, coastal dunes, desert coast, Patagonian steppe, low bushes and highland grassland. **Voice** Variation suggests different species. In central Chile repetitive calls given from the ground (*Kew-kew-kew-kew-...*), but in Patagonia a short phrase of two syllables, strident and repeated 4–10 times, sometimes ending in a melodic trill (*Weet-it-weet-it- weet-it- weet-it -... trrrrrrrrrr...*). **Where to see** Las Cuevas/Chungará road (Arica y Parinacota Region), Los Choros gorge (Coquimbo Region), Las Cruces/Punta de Tralca (Valparaíso Region), Pichilemu (O'Higgins Region), Carelmapu (Los Lagos Region), Otway Sound (Magallanes Region), Sierra Baguales (Magallanes Region).

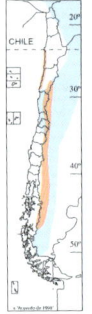

Rufous-banded Miner
Geositta rufipennis R
L 16–18cm. Sexes identical. 0–3,000m. **ID** Similar to other miners. In fresh plumage, rufous tone to flanks. Bill rather straight and short. In flight, primaries, secondaries and rectrices are intense orange, with black terminal band to tail. Inhabits Andes in spring/summer. In the north, coastal rocky areas with scattered vegetation and Andean foothills in winter. In the south, at edge of forest (*Nothofagus* spp.). **Voice** In spring and summer its song is characteristic of central Chilean Andes: two parts, both repeated, with first becoming longer with each repetition; high-pitched, loud and shrill, the first part scratchy and the second a trill (*Trrrrrr...Tchet-tchet-tchet-tchet-tchet-tchet-...Trrrrree-Tchet-tchet-tchet...*), gradually decreasing in volume. Some local dialects, but overall variation throughout country is rather small. **Where to see** Carrizal Bajo village (Atacama Region), Punta de Choros village (Coquimbo Region), Farellones, El Yeso Valley (Metropolitan Region), Termas del Flaco (O'Higgins Region).

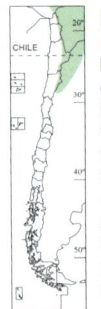

Puna Miner
Geositta punensis R
L 14–15cm. Sexes identical. >4,000m. **ID** Similar to other miners. Whitish breast and belly, otherwise pale brown. Bill somewhat curved. Perches very upright. In flight, primaries and secondaries orange-cinnamon, with black terminal band to pale orange-cinnamon tail. **Habitat** Altiplano waterbodies, pampas, and rocky slopes. Shuns waterlogged areas and high-Andean bogs (bofedales). **Voice** Most frequently heard is a monosyllabic contact call, given from the ground or in flight, a rather short dry *piew!* Song is elaborate and complex, sometimes begins with a quick repetition of the contact call, followed by a long sequence of trills, reminiscent of the song of Cordilleran Canastero. **Where to see** Parinacota, Las Cuevas/Chungará road, Chungará lake and Tambo Quemado pass (Arica y Parinacota Region). Tara Salt Flat, Miscanti and Miñique lagoons (Antofagasta Region).

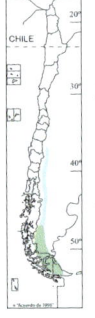

Short-billed Miner
Geositta antarctica R
L 15–16cm. Sexes identical. 0–200m. **ID** Similar to other miners. Dirty grey-white breast, scaled rather than streaked. Short bill. At rest wings almost as long as tail, with long primaries. In flight, pale brown wings and black tail with broad white border. **Habitat** Open Patagonian steppe and sandy areas near coast. Low scrub (*Berberis* sp., *Baccharis* sp.) but not forest. **Voice** Limited repertoire. Most commonly heard are: a sharp, scratchy trill, repeated several times in quick succession (*trrrrr-trrrrr-trrrr –trrrrr-trrrr-trrrr...*), and a clear, energetic, short dry chirp, sometimes repeated at long intervals (*pit*). **Where to see** Pali Ayke National Park, Pingüino Rey National Park, Porvenir (Magallanes Region).

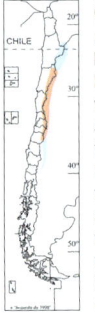

Creamy-rumped Miner
Geositta isabellina R
L 17–19cm. Sexes identical. 2,700–4,000m. **ID** Similar to other miners. Dirty grey breast. At rest wings almost as long as tail. In flight, cinnamon wings and mainly white tail (and rump) has dark brown terminal band. **Habitat** High Andean environments. It prefers stony slopes near water. **Voice** Male very vocal when breeding. Song lasts c.10 seconds and comprises 40–50 notes, overall scratchy and with a somewhat unmistakable metallic ring. Starts with c.4–5 notes at a given volume, which then increase, before declining in loudness (*Trrrrrwet-trrrrrwet-trrrrrwet-trrrrrwet- Trrrrwet- Trrrrwet-Trrrwet-Trrrrwet-Trrrrwet-trrrrwet-trrrrrwet-trrrwet-trrwet...*). **Where to see** Tara Salt Flat (Antofagasta Region), Portillo (Valparaíso Region), Nevado and El Yeso Valleys (Metropolitan Region), Rancagua Andes (O'Higgins Region).

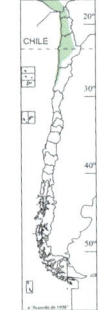

Greyish Miner
Geositta maritima R
L 14–15cm. Sexes identical. 0–3,500m. **ID** Similar to other miners and ground-tyrants. Overall greyish-brown. Typical posture rather horizontal than upright. In flight wings and tail blackish-brown. **Habitat** Extremely arid environments, coastal or inland. **Voice** A clear, dry and irregularly repeated note, changing slightly in tone and volume, apparently in contact (*Piew-piew-Piew-piew ...*). Also a short sharp trill or rattle. **Where to see** Arica–Putre road (Arica y Parinacota Region), Antofagasta–Tal Tal road (Antofagasta Region), San Pedro de Atacama (Antofagasta Region), road to Totoralillo (Atacama Region).

PLATE 68: EARTHCREEPERS

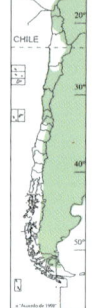

Scale-throated Earthcreeper
Upucerthia dumetaria   **R**

L 21–22cm. Sexes identical. 0–4,000m. **ID** Similar to Patagonian Forest Earthcreeper. Generally pale coloration. Pale superciliary. Long bill. Tail with outer edges paler and warmer than rest. **Habitat** Arid and semi-arid open environments, steppe, and low vegetation. **Voice** Song is a repeated series of 11–13 notes that increases in intensity (*twik - twik-twik - twik - twik - twiktwiktwik...*), which is heard everywhere in the Andes of central Chile. **Where to see** Pica/Matilla (Tarapacá Region), Tamarugo forest near San Pedro/Toconao (Antofagasta Region), Parque Andino Juncal (Valparaíso Region), Farellones, El Yeso Valley (Metropolitan Region), Sierra Baguales (Magallanes Region).

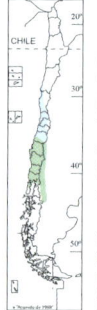

Patagonian Forest Earthcreeper
Upucerthia saturatior   **R**

L 15–16cm. Sexes identical. 0–2,000m. **ID** Similar to Scale-throated Earthcreeper. Generally dark. Long and well-marked superciliary. Bill not as long. Tail has brown-rufous outer edges. **Habitat** Forest and undergrowth, mainly *Nothofagus* sp. In winter in central Chile also mountains, gorges and coastal cliffs. **Voice** Song rendered *p'p-tirik-tirik-tiriktirik-tiruk...*, with the *tirik* note recalling a parrot. Very different from Scale-throated Earthcreeper. **Where to see** Parque Andino Aguas de Ramón, Parque Panul (Metropolitan Region), Los Cipreses National Park (O'Higgins Region), Altos de Lircay National Reserve (Maule Region), Termas de Chillán (Ñuble Region). **Note** Until recently, considered a subspecies of Scale-throated Earthcreeper.

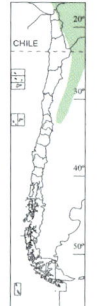

Buff-breasted Earthcreeper
Upucerthia validirostris   **R**

L 19–20cm. Sexes identical. 3,000–4,600m. **ID** Similar to White-throated Earthcreeper. Pale brown, with a slightly marked supercilium and usually an unmarked throat. Breast and belly uniform pale brown. **Habitat** Andean slopes, in scrubby gorges, and in highlands, in grassland (*Festuca* sp.) and rocky slopes with low vegetation. **Voice** Song is a very long, high-pitched trill, with a metallic quality and an even tempo throughout, although it sometimes commences with 3–4 clearly separated notes (*Kip ... Kip ... Kip ... Kip ... Kipkipkipkipkipkipkipkipkipkipkip...*). **Where to see** Putre and Chungará lake (Arica y Parinacota Region). **Note** Recent work has concluded that Plain-breasted Earthcreeper *U. jelskii* and *U. validirostris* represent a single species, with *jelskii* relegated to subspecies level.

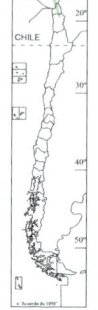

White-throated Earthcreeper
Upucerthia albigula   **R**

L 20–22cm. Sexes identical. 3,000–3,800m. **ID** Similar to Buff-breasted Earthcreeper. Small white throat, dark-marked ear-coverts, and generally rufous, warm coloration. Sometimes shows marks on throat and upper breast in fresh plumage. **Habitat** Andean foothills, in gorges with shrubby vegetation. **Voice** Song is long and almost strident, with the same note repeated slowly and certainly at the start, then slightly accelerating and becoming higher-pitched (*Pew ... Pew... Pew ... Pew ... Piúpiupiupiupiupiupiupiu...*). **Where to see** Andean foothills in the Arica y Parinacota Region, e.g. Socoroma, Putre, Chapiquiña, etc.

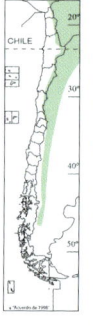

Straight-billed Earthcreeper
Ochetorhynchus ruficaudus   **R**

L 18–19cm. Sexes identical. **N** 2,000–4,600m. **C** 2,500–3,500m. **ID** Similar to Cordilleran Canastero. Rufous back, white breast, and runs quickly on the ground, with raised tail and long, slightly curved bill. **Habitat** High Andes, being common in rocky areas with bushes in foothills in northern and central Chile, but also reaches down to limit of desert in the north. **Voice** Audible at some distance, a very sharp and penetrating monosyllable *Peeep!*, heard very infrequently. Song is a very sharp trill, starting higher, becoming slightly squeaky and then decreasing in volume (*Peep-peep-peep-Peep-PEEP-PEEP-PEEP-Peep-peep-peep...*). **Where to see** Socoroma/Putre (Arica y Parinacota Region), around Machuca (Antofagasta Region).

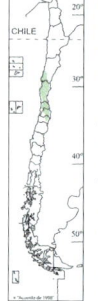

Crag Chilia
Ochetorhynchus melanurus   **Re**

L 17–19cm. Sexes identical. 0–3,000m. **ID** Unmistakable. Note combination of rufous body and white breast, unique in its habitat. **Habitat** Andean foothills, sparse sclerophyll vegetation or succulent cover. Shuns forested areas, but in hard winters will seek denser vegetation on lower slopes. **Voice** Song lasts c.2 seconds, and is strident, initially increasing in intensity, then gradually decreasing, and usually starts with a drawn-out growl (*Krrrrrrrr-tee-teee-teeee-teeee*). In probable alarm, constantly repeats a series of four notes (*tchip-tchip-tchip...tchip...*). **Where to see** Santa Gracia (Coquimbo Region), La Campana National Park (Valparaíso Region), El Yeso Reservoir road (Metropolitan Region).

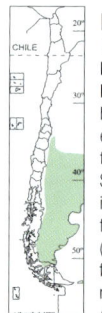

Band-tailed Earthcreeper
Ochetorhynchus phoenicurus   **R**

L 16–18cm. Sexes identical. **A** <300m. **ID** Unmistakable. Brown above with rufous-based black tail, long white supercilium, slightly rufous ear-coverts, and pale throat. **Habitat** Steppes and thickets in Patagonia. **Voice** Simple repertoire. Song is a sharp trill that lasts 2–3 seconds, more intense in the middle and then declines, sometimes followed by three short, sharp accentuated notes (*crrrrrrRRRRRrrrrr ... Pee- Pee- Pee ...*). Song given from perch in bush, is sometimes accompanied by raised wings, like a cincloides. **Where to see** Primera Angostura, Sierra Baguales (Magallanes Region).

PLATE 69: CANASTEROS

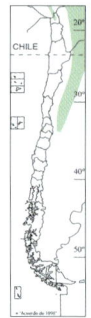

Arequipa Canastero
Asthenes arequipae   R
L 15–17cm. Sexes identical. **N** 3,000–4,000m.
ID Similar to Canyon Canastero. Contrasting black, brown and rufous coloration, and robust bill. Chooses exposed perches. **Habitat** Semi-arid foothills with shrubs and cacti. Very common in Queñoa (*Polylepis rugulosa*) forests in Tarapacá. **Voice** Song starts with 2–3 low, soft notes, then becomes a high-pitched trill that slightly rises and falls, and lasts c.3 seconds (*Pee-Peepee-Trrrrrrrrrriiiip....*), commonly given from atop a bush. **Where to see** Arica y Parinacota Region, e.g. Socoroma, Putre or Ticnamar. **Note** Formerly treated as conspecific with Creamy-breasted Canastero *A. dorbignyi*. [Alt. Dark-winged Canastero]

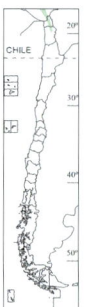

Canyon Canastero
Asthenes pudibunda   R
L 16–17cm. Sexes identical. **N** 3,000–3,700m.
ID Similar to Creamy-breasted Canastero. A rather rufous canastero, which rarely perches in open. Elusive, moving between bushes in low flights. **Habitat** Semi-arid foothills of Andes; well-vegetated gulleys and valleys. **Voice** Song lasts c.3 seconds and with 3–4 separate notes that accelerate into a stable trill of 30–40 notes, and which is repeated every 4–5 seconds (*peip-peip-peip-peip-pipipipipipipipipipipip ...*). **Where to see** Socoroma and Putre (Arica y Parinacota Region).

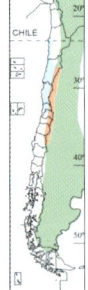

Cordilleran Canastero
Asthenes modesta   R
L 14–16cm. Sexes identical. **N** 3,000–4,600m.
C 0–3,000m. **A** <1,000m. **ID** Similar to Sharp-billed Canastero and Straight-billed Earthcreeper. Well-patterned, rufous-and-black tail. Terrestrial, periodically using an exposed perch. **Habitat** Rocky areas. In the north, pre-puna with sparse thickets and altiplano steppe. In central Chile, mountains with low scrub in spring and summer. In the south, Patagonian steppe with low scrub. **Voice** Commonly heard in Andes. Song is a very sharp trill of 20–25 notes (inseparable to the ear, except first three) that lasts c.1.5 seconds and ascends slightly (*twe-twe-twe-twe-twe-twe-twetwetwetweeerrrrrr...*). This is repeated 5–6 times, at intervals of 1.0–1.5 seconds. **Where to see** Parinacota (Arica y Parinacota Region), Farellones, El Yeso Valley (Metropolitan Region), Pali Ayke National Park (Magallanes Region), coasts in winter (Antofagasta Region–Coquimbo Region).

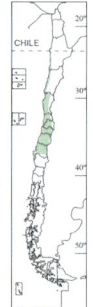

Dusky-tailed Canastero
Pseudasthenes humicola   Re
L 15–17cm. Sexes identical. **C** 0–2,200m.
ID Could be confused with Plain-mantled Tit-Spinetail, but tail is dark and not especially long. **Habitat** Spiny thickets and sclerophyllous forest. **Voice** Very vocal when nesting. Song lasts 2–3 seconds and is repeated every 5–10 seconds; it can contain more than 30 notes, given in pairs that start slowly and accelerate slightly, without ever becoming trilled (*P'tew..P'tew..P'tew..P'tew..P'tew..p'tew-p'tew-p'tew-p'tew-p'tew-p'tew-p'tew...*). Also a very peculiar distorted, nasal and metallic *chwei-chwei-chwei-chwei-chwei-chwei-chwei-choo...*, with shorter variations. Also a long trill of 3–5 seconds, with slight oscillations, rising and falling. **Where to see** Parque Mahuida, Farellones road (Metropolitan Region), coastal range (Coquimbo Region–Maule Region), ravines and coastal plateaus (Coquimbo Region–Bio Bio Region).

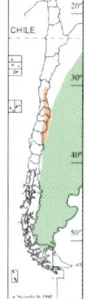

Sharp-billed Canastero
Asthenes pyrrholeuca   R
L 15–17cm. Sexes identical. **C** 0–3,000m. **S & A** 0–1,500. **ID** In central and southern Chile could be confused with Cordilleran Canastero. Long dark tail, and sharp bill. Keeps permanently inside bushes. **Habitat** Tangled scrub. In central Chile, in mountains during spring and summer. Only in Magellan Region does it reach sea level. **Voice** A short and characteristic *tweet* in contact, which can be repeated and slightly louder in alarm (*Tweet-Tweet ...Tweet-Tweet...*). Song easily heard in its environment, comprising three short and two longer notes (*Tweet-Tweet -Tweet – tueititit – tueititit...*). Also a trill that recalls other members of the genus. **Where to see** Farellones, El Yeso Valley (Metropolitan Region), Pali Ayke National Park (Arica y Parinacota Region).

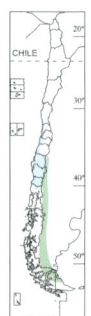

Austral Canastero
Asthenes anthoides   R
L 16–17cm. Sexes identical. **A** 0–1,500m.
ID Compare Correndera Pipit. Blackish upperparts with pale fringes, no malar or dark markings below. **Habitat** Bushy steppe and Patagonian scrub. Never in dense forest or grassland. **Voice** Small repertoire. Song is a long high-pitched trill, sometimes slightly ascending and occasionally preceded by two short notes, but usually a simple trill (*trrrruuuuiiiiiuit*). Often heard in its habitat. **Where to see** Concón and Viña del Mar in winter (Valparaíso Region), Pali Ayke National Park, Porvenir (Magallanes Region).

PLATE 70: TIT-SPINETAILS AND TIT-TYRANTS

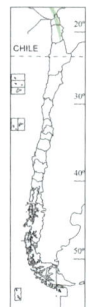

Streaked Tit-Spinetail
Leptasthenura striata R
L 16–17cm. Sexes identical. **N** 1,500–3,800m.
ID Similar to Puna Tit-Spinetail. Body covered with obvious pale streaks, especially on mantle and breast. **Habitat** Semi-arid areas, scrub and cacti. Also Queñoa forests (*Polylepis* spp.). **Voice** Mostly silent. Song repeats a short, dry, drawn-out, scratchy note (*tchek… tchek-tchek… tcheck…*). **Where to see** Putre and Socoroma (Arica y Parinacota Region).

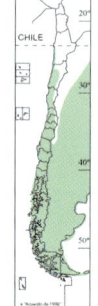

Tufted Tit-Tyrant
Anairetes parulus R
L 10–11cm. Sexes similar. **C** 0–2,000m.
ID Unmistakable in Chile. Very small, and almost always in pairs. **Habitat** Thickets and arid slopes; also sclerophyll forest. **Voice** Pairs maintain continuous contact with short high-pitched whistles (*tit… tit…*), sometimes disyllabic (*Tchi-wit … Tchi-wit…*). Probably in territory advertisement gives a short trill, lasting c.1 second, every 5–6 seconds, without any change in pitch (*ptrrrrrrrrrrrrrrrrrrrrrrrrrrrrrt…*). **Where to see** El Peral lagoon (Valparaíso Region), Parque Mahuida (Metropolitan Region). Fairly common.

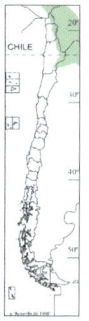

Puna Tit-Spinetail
Leptasthenura berlepschi R
L 16–17cm. Sexes identical. **N** 2,500–4,600m.
ID Could be confused with Streaked Tit-Spinetail. Back and belly warm-coloured, without streaks above, and has whitish superciliary reaching towards nape. **Habitat** Open areas, thickets and *Polylepis* forests, even steppes. **Voice** Calls constantly. A short trill (*trrrrrrr*) in contact. Lasting c.2 seconds, with c.20 notes is a very metallic *tik-tik-tik-tik-tik-tik-tik…*. **Where to see** Putre/Socoroma (Arica y Parinacota Region), San Pedro de Atacama and Peine (Antofagasta Region). **Note** Formerly treated as conspecific with Plain-mantled Tit-Spinetail. [Alt. Buffy Tit-Spinetail]

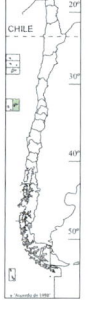

Juan Fernandez Tit-Tyrant
Anairetes fernandezianus Re
L 12–14cm. Sexes identical. 0–300m.
ID Unmistakable. Unique tit-tyrant on Juan Fernandez archipelago, with typical plumage and tall spiky crest. **Habitat** Woodland and scrubby areas, especially native forest of *Myrceugenia fernandeziana*. **Voice** Song is a sustained note, initially quiet that gradually increases in volume so that, after 5–6 seconds, it becomes a long trill, the whole eventually lasting 10–15 seconds (*grrrrrrrrrrrrrrrr-trewtrewtrewtrewtrewtrew …*). **Where to see** Robinson Crusoe Island (or Más a Tierra) in Juan Fernández archipelago (Valparaíso Region).

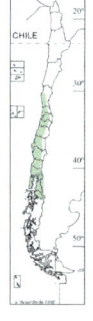

Plain-mantled Tit-Spinetail
Leptasthenura aegithaloides R
L 16–17cm. Sexes identical. **C** 0–2,600m. **S & A** 0–1,500m. **ID** Unmistakable. Similar to other tit-spinetails in Chile, but there is no overlap – it is the only bird with this kind of tail in its range. Ssp. *berlepschi* has mantle and belly unstreaked, and has white superciliary reaching towards neck. **Habitat** Arid areas, thickets and forests. Also parks and urban areas. **Voice** Calls constantly. In contact, a three-note, scratchy call that lasts c.0.5 seconds and recalls an insect; it starts strongly then fades (*Krrggrrr…Krrggrr…*) and is given frequently while foraging or in flight. Song comprises 9–10 very high notes, the first drawn-out and the rest even, and in whole lasting little more than c.1 second (*Fii-fii-fii-fii-fiit… Fii-fii-fii-fii-fii-fii-fii-fiit…*). **Where to see** Metropolitan Park (Metropolitan Region), coastal range (Coquimbo Region–Bio-Bio Region), etc.

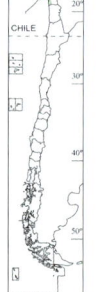

Pied-crested Tit-Tyrant
Anairetes reguloides R
L 11–12cm. Sexually dimorphic. 0–1,000m.
ID Unmistakable. Dark or black face, with two pointed crests, resembling horns, and a white crown. **Habitat** Ravines with bushy vegetation and, marginally, in planted fields. **Voice** Song of 6–8 clearly separated notes is slightly descending, with the first a rather sharp whistle and the rest repeated (*peeep-tweet-tweet - tweet- tweet - tweet- tweet…*), usually lasting 2–3 seconds. In variation, after first 4–5 notes sometimes gives a long harsh trill, lasting 4–5 seconds. **Where to see** Very local in Chile. Camarones Valley (Arica y Parinacota Region).

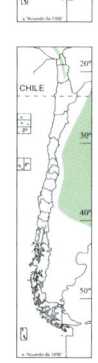

Yellow-billed Tit-Tyrant
Anairetes flavirostris R
L 10–12cm. Sexes similar. **N** 1,200–3,600m.
ID Unmistakable in Chile. Dark iris, orange-yellow base to lower mandible and heavy streaks on breast. **Habitat** Andean foothills, in wooded areas and fields. **Voice** Pair members regularly give contact calls, a short, rather low trill, lasting less than 1 second, recalling a cricket (*krrrrrrrrrrii*). Song varies, but usually starts with four notes which sound like two to the human ear, and continues as a long trill (*petew-trrrrrrrrrrrrrrrrrrrrrrrrrr*). **Where to see** Foothills in the Arica y Parinacota Region, e.g. Socoroma and Putre.

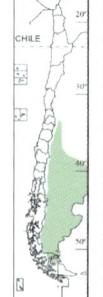

Patagonian Tit-Spinetail
Leptasthenura pallida R
L 17–18cm. Sexes identical. **S & A** 0–1,200m.
ID Unmistakeable. Similar to other tit-spinetails in Chile, but no overlap in its range. Body mostly greyish with cinnamon crown. Upperparts and belly greyish, with pale streaks above. Tail longer than other tit-spinetails; also has the shortest bill. **Habitat** Bushy steppe, Patagonian scrub and forest edges. Also in semi-arid areas with scattered bushes. **Voice** Calls constantly, while foraging and moving in bushes (each call lasting 3–4 seconds). **Where to see** Road from Puerto Natales to Torres del Paine National Park, Sierra Baguales and bushy areas near Porvenir (Magallanes Region).

PLATE 71: RAYADITOS, WIRETAIL, RUSHBIRD, DORADITO AND RUSH-TYRANT

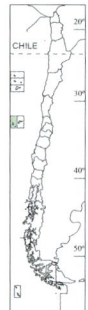

Masafuera Rayadito
Aphrastura masafuerae **Re**
L 15–17cm. Sexes identical. 700–1,300m.
ID Unmistakable. Plumage very similar to mainland Thorn-tailed Rayadito, but unique within the species' insular range. **Habitat** Dense vegetation. Strongly associated with tree-fern forests (*Dicksonia* sp.) or areas with ferns (*Lophosoria* sp.). **Voice** No information. **Where to see** Only on Alejandro Selkirk Island in Juan Fernández archipelago (Valparaíso Region).

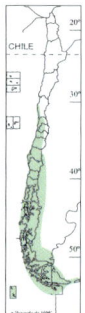

Thorn-tailed Rayadito
Aphrastura spinicauda **R**
L 14–15cm. Sexes identical. **C, S & A** 0–2,400m.
ID Unmistakable. Small, fast, relentless and noisy bird that is constantly jumping between branches. Very bold buff supercilium on otherwise black head, and broad buff markings on dark wings. Unusual tail structure, with retrices ending in rachis (not barbs), like thorns, used as support. **Habitat** Wet sclerophyll and temperate forests. In the south in scrub and even Patagonian steppes covered with *Festuca* sp. **Voice** Five types of vocalisations have been described. When alert gives a single or several monosyllables, which accelerate (c.6+ notes per second), and is continuously uttered until the threat disappears (*tsit-tsit-tsit-tsit-tsit-tsit-tsit-tsittsit-tsit-tsit-tsit-tsit-tsit...*). Song is a repeated trill, sometimes given by both pair members alternately, c.5–6 times per 30 seconds (*Triiiiiiiet-Triiiiiiiet-Triiiiiiiet...*). **Where to see** Fray Jorge National Park (Coquimbo Region), Zapallar, Córdoba gorge, La Campana National Park (Valparaíso Region), Nahuelbuta National Park, Cerro Ñielol (Araucanía Region), Chiloé Island (Los Lagos Region), Fuerte Bulnes road (Magallanes Region).

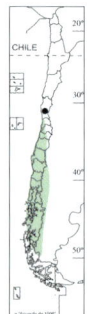

Des Murs's Wiretail
Sylviorthorhynchus desmursii **R**
L 22–24cm. Sexes identical. **A** 0–1,000m.
ID Unmistakable but birds without a full tail could be confused with Southern House Wren. Note orange crown, sharp-pointed, straight bill, and unbarred wings. **Habitat** Undergrowth. At northern limit in sclerophyllous forests (*Peumus* sp., *Quillaja* sp. and *Lithrea* sp.). In the south in forests of *Nothofagus* and areas of native bamboo (*Chusquea* sp.) or Espinillo (*Ulex europaeus*). **Voice** Very vocal. Proclaims territory with a high-pitched phrase repeated 3–5 times (*T'piyee-T'piyee-T'piyee-T'piyee-T'piyee...*), then again after a few seconds. In alarm, a long trill repeated every c.5 seconds, *prrrrrrrieeeep... prrrrrrrieeeep... prrrrrrrieeeep....* Song complex, varied and often starts with a repeated note, then becomes high-pitched and excited. **Where to see** Córdoba gorge, La Campana National Park (Valparaíso Region), Conguillio National Park (Araucanía Region), Nahuelbuta National Park, Cerro Ñielol (Araucanía Region), Chiloé Island (Los Lagos Region).

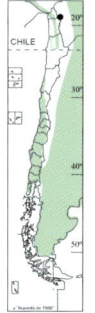

Wren-like Rushbird
Phleocryptes melanops **R**
L 13–15cm. Sexes identical. **N** 0–2,500m.
C 0–1,500m. **S & A** 0–300m. **ID** Unmistakable. Common to see this bird very briefly, and more often heard than seen. With a compact body and elongated bill, its behviaour makes it difficult to see details of its coloration. Rufous-orange wing-stripes distinguish it when flying between reeds. Greyish dorsal lines are only visible when it moves slowly or in favourable light. **Habitat** Reeds (*Scirpus* sp. and *Typha* sp.) and rushes beside standing and flowing water. **Voice** Calls frequently. In contact, high-pitched and scratchy monosyllables, variably repeated (*Trrek... Trrek... Trrek...*) and recalling a tree in the wind. Another monosyllable (*krrrrc... krrrrc...*), reminiscent of a motor, appears to signify alertness and can precede the song, which recalls two stones being hit, or a small two-stroke engine, with 7–9 'strokes' per second, but no changes in volume or intonation (*tiktiktiktiktiktiktiktiktik*). **Where to see** Loa River (Antofagasta Region), El Peral lagoon (Valparaíso Region), Lampa/Batuco system (Metropolitan Region), Torca lagoon (Maule Region), etc. All wetlands.

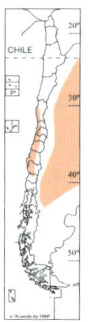

Ticking Doradito
Pseudocolopteryx citreola **R**
L 11–13cm. Sexes identical. 0–500m.
ID Unmistakable in Chile. Yellow underparts from chin to undertail. **Habitat** Dense scrubby vegetation around wetlands, but also in reeds (*Scirpus* sp., *Typha* sp.) **Voice** Song, long and scratchy, somewhat devoid of musicality, lasting c.4 seconds, comprises seven syllables, four introductory, three terminal, the latter slightly louder (*tchoek-tchoek-tchoek-tchoek-Kek-Kerek-Kerew ...*). Sometimes only the introductory notes are given, with a more relaxed rhythm, and 1.5-second pauses between them. **Where to see** Rio Maipo wetland (Valparaíso Region), Lampa/Batuco system (Metropolitan Region), Petrel lagoon (O'Higgins Region), Putú wetland (Maule Region), Purén wetland (Araucanía Region).

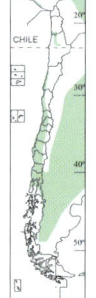

Many-coloured Rush-Tyrant
Tachuris rubrigastra **R**
L 10–12cm. Sexes identical. **N** 0–2,500m.
C & S 0–800m. **ID** Unmistakable. A stunning jewel-like bird, with largely green upperparts and yellow underparts, blue face, red central crown and vent, black breast-band, bold white slash across wings and long pale yellow supercilium. There is no other similar-looking bird found in the reeds environment. **Habitat** Reeds (*Scirpus* sp., *Typha* sp.) and other vegetation around wetlands, estuaries and irrigation canals. **Voice** Calls regularly. In contact, a sharp, monosyllable at irregular intervals, *piew*. In alarm, a high-pitched note, repeated regularly, *Pet'ew... Pet'ew... Pet'ew....* Song is a long phrase, lasting c.2.5 seconds, comprising ten syllables separated into two parts, and repeated every 4–5 seconds (*Pet'ew-Kha... Pet'ew - pet'ew - pet'ew - pet'ew - pet'ew -tew-tew-tew...*). **Where to see** Loa River (Antofagasta Region), Rio Maipo wetland, El Peral lagoon (Valparaíso Region), Lampa/Batuco system (Metropolitan Region), Petrel lagoon (O'Higgins Region), Budi lake (Araucanía Region), Chiloé Island (Los Lagos Region).

PLATE 72: NEGRITOS, TYRANT AND GROUND-TYRANTS I

Austral Negrito
Lessonia rufa   R
L 11–13cm. Sexually dimorphic. 0–2,600m. **ID** Male like male Andean Negrito. Compare female to Spot-billed Ground Tyrant. Male has rufous mantle and black primaries on both surfaces. Female is pale brown with rufous-ochre mantle. Dark wings and tail, and all-black bill. **Habitat** Around wetlands including estuaries, lakes, lagoons and coasts. **Voice** Quiet and only audible close to. In contact, a very high, quiet monosyllable, singly or repeated (*peet*). Also a short trill comprising three or four notes, inseparable to the ear (*petriee*), repeated at intervals. **Where to see** Rio Maipo wetland (Valparaíso Region), Lampa/Batuco system (Metropolitan Region), Chiloé Island (Los Lagos Region).

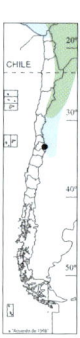

Andean Negrito
Lessonia oreas   R
L 12–13cm. Sexually dimorphic. 0–4,700m. **ID** Similar to male Austral Negrito. Females of the two species are rather different. Male has ochre mantle and whitish primaries on both surfaces. **Habitat** Around high-Andean bogs (bofedales), salt flats, estuaries, lakeshore and even coasts. **Voice** In contact, a high-pitched quiet *peet*. Male in display flies diagonally upwards, turns sharply and returns to the ground, producing a mechanical sound with the wings. **Where to see** Parinacota and Chungará lake (Arica y Parinacota Region), Huasco Salt Flat (Tarapacá Region), Machuca and Atacama Salt Flats (Antofagasta Region).

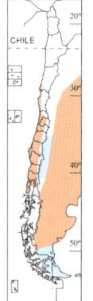

Spectacled Tyrant
Hymenops perspicillatus   R
L 15–16cm. Sexually dimorphic. 0–1,200m. **ID** Male unmistakable. Female has rufous or more orange wings, very evident in flight, and yellow eye-ring. **Habitat** Around wetlands, estuaries, lakes and lagoons. **Voice** Mechanical sounds and quiet calls. Male display flight reaches c.5m above the ground and describes an elliptical path, emitting a buzz during the descent, apparently using the primaries, and a high *click* at the apex. Male also calls sporadically from a perch, a quiet, short, high-pitched whistle followed by a crunching sound (*peee-tchrrk*). Female gives a pitiful or rather squeaky *twee*, sometimes repeated. **Where to see** Elqui River (Coquimbo Region), Rio Maipo wetland (Valparaíso Region), Lampa/Batuco system (Metropolitan Region).

GROUND-TYRANTS Very similar insectivores, of delicate appearance and shared behaviours. Fine bills, well-proportioned, long legs and highly terrestrial. Crown colour very important for identification.

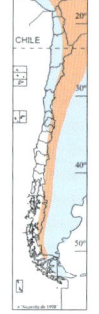

Spot-billed Ground-Tyrant
Muscisaxicola maculirostris   R
L 14–15cm. Sexes identical. 0–4,000m. **ID** Similar to other ground-tyrants and female Austral Negrito. Small with a proportionately large head and orangey bill base. Two wingbars in fresh plumage, lost when worn. **Habitat** Semi-arid, stony areas and very dry vegetation. Also Queñoa *Polylepis* sp. forests, Tamarugo *Prosopis* sp. forests, even salt flats. **Voice** The most frequently heard ground-tyrant. A short, dry and very quiet contact call, audible only when very close, given once or repeated at irregular intervals (*pip*). Song louder, given during display flight and has two parts, repeated every 2–3 seconds. The introduction comprises 3–4 syllables, sometimes just 2–3, followed by three syllables inseparable to the human ear (*pip-pip-pip-PipPiriWeet… pip-pip-pip- PipPiriWeet …*). **Where to see** Putre in winter (Arica y Parinacota Region), Llanos de Challe National Park (Atacama Region), Farellones and El Yeso Valley (Metropolitan Region).

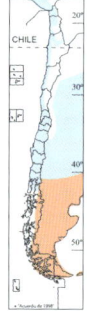

Dark-faced Ground-Tyrant
Muscisaxicola maclovianus   R
L 15–16cm. Sexes identical. **N** 0–4,600m. **C** 0–3,500m. **S & A** 0–1,500m. **ID** Similar to Cinnamon-bellied Ground-Tyrant. Smoky-black face and grey body. **Habitat** Coasts to mountains, in open country or with low vegetation. On migration anywhere, even in cities. **Voice** Most frequently heard is a high, dry monosyllabic whistle, *f'eet*, audible only at close range, which may become a series during aggressive interactions, some of them longer, but apparently without definite pattern. **Where to see** In winter, common on central coast, e.g. Punta Teatinos (Coquimbo Region), Viña del Mar or Cachagua (Valparaíso Region). In spring, on road to Fuerte Bulnes and near Pali Ayke National Park (Magallanes Region).

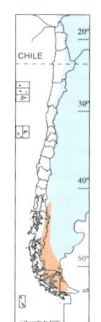

Cinnamon-bellied Ground-Tyrant
Muscisaxicola capistratus   R
L 17–18cm. Sexes identical. **A** 0–1,500m. **ID** Similar to Dark-faced Ground-Tyrant. Smoky-black face and warm cinnamon underparts, especially flanks and belly. Intense rufous crown. **Habitat** Patagonian steppe. Grassland with bushes. High-Andean steppe on migration. **Voice** A high-pitched, monosyllabic, quiet whistle (*fit*), sometimes mixed with short trills lasting several seconds (*fit… fit… treet… treet… treeet fit…*). **Where to see** In autumn and early spring, through central and northern mountains, to 4,500m, usually alone. In spring and summer, Pali Ayke National Park and Sierra Baguales (Magallanes Region).

154

PLATE 73: GROUND-TYRANTS II

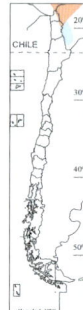

White-fronted Ground-Tyrant
Muscisaxicola albifrons   R
L 22–24cm. Sexes identical. >4,000m. **ID** Unmistakable. Large and erect, with distinctly rounded white forehead and diffuse reddish-brown crown patch. Wings with notable pale edges in secondaries and coverts. **Habitat** Altiplano, in high-Andean bogs (bofedales) and their immediate surroundings. **Voice** A short, monosyllabic, scratchy *pieuk*, apparently used in both intraspecific aggression and contact between pair members. **Where to see** Las Cuevas, Parinacota and Guallatire bofedales (Arica y Parinacota Region).

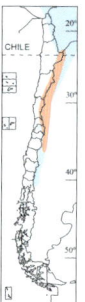

Black-fronted Ground-Tyrant
Muscisaxicola frontalis   R
L 19–20cm. Sexes identical. **N** 3,500–4,500m. **C** 2,500–3,500m. **ID** Similar to other ground-tyrants. Upright posture, with extensive black on forehead and crown. **Habitat** Montane stony areas on slopes and near water. **Voice** Gives short, quiet, high-pitched whistles similar to congeners (*fit*). **Where to see** El Juncal Andean Park (Valparaíso Region), Farellones and El Yeso Valley (Metropolitan Region), Termas de Chillán (Ñuble Region).

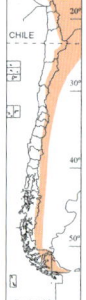

Ochre-naped Ground-Tyrant
Muscisaxicola flavinucha   R
L 18–20cm. Sexes identical. **N** >4,000m. **C** 2,000–3,500m. **S & A** 0–1,000m. **ID** Unmistakable, by yellowish patch on crown with well-marked white forehead. **Habitat** High-Andean bogs (bofedales) and surrounding altiplano; slopes, grassland and Andean valleys near water in central Chile, and steppes in the south. **Voice** A short, repeated *pip-pip-pip*, and a very high *feeet*; in both cases their context is unknown. **Where to see** Las Cuevas and Parinacota bofedales (Arica y Parinacota Region), Quepiaco bofedales (Antofagasta Region), El Juncal Andean Park (Valparaíso Region), Farellones and El Yeso Valley (Metropolitan Region).

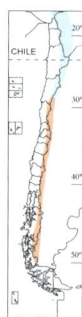

White-browed Ground-Tyrant
Muscisaxicola albilora   R
L 17–18cm. Sexes identical. **C** 1,800–3,000m. **S & A** 1,000–3,000m. **ID** Similar to other ground-tyrants. Greyish-brown mantle with a diffuse rufous patch on crown and obvious white eyebrow. **Habitat** Rocky slopes, scrubby areas and grassland in mountains. **Voice** Most frequently heard is a quiet monosyllabic whistle (*fit*), which sometimes becomes a sequence and can sound drawn-out in aggressive interactions, when also utters variable mouse-like grunts (*quiq... quiq...*) accompanied by raised wings. **Where to see** El Juncal Andean Park (Valparaíso Region), El Yeso Valley (Metropolitan Region), Termas de Chillán (Ñuble Region).

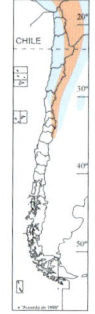

Rufous-naped Ground-Tyrant
Muscisaxicola rufivertex   R
L 16–17cm. Sexes identical. **N** 0–4,600m. **C** 2,000–3,500m. **ID** Similar to other ground-tyrants. Pale grey mantle, dark wings, and well-defined red crown patch. Bill has slight hook. **Habitat** Prefers rather dry areas with low vegetation in mountains. In the north, common near salt flats, and also on coasts. **Voice** Two types of short, probably contact calls, which can be repeated, or given together; both are monosyllabic, high whistles (*feet... fuit...*). **Where to see** Surire Salt Flat (Arica y Parinacota Region), Pampa del Tamarugal National Reserve (Tarapacá Region), near Toconao and Atacama Salt Flat (Antofagasta Region), El Juncal Andean Park (Valparaíso Region), road to Valle Nevado (Metropolitan Region).

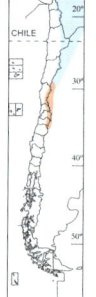

Cinereous Ground-Tyrant
Muscisaxicola cinereus   R
L 16–17cm. Sexes identical. >2,300m. **ID** Similar to other ground-tyrants. Grey upperparts with concolorous crown, which is unique among Chile's ground-tyrants. Fine, straight bill. **Habitat** Mountains, usually above 3,000m. Rocky areas near water. **Voice** A soft, brief *peep*, repeated irregularly. **Where to see** El Yeso Valley and upper Maipo canyon (Metropolitan Region).

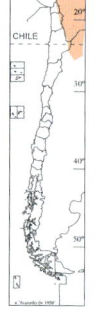

Puna Ground-Tyrant
Muscisaxicola juninensis   R
L 16–17cm. Sexes identical. >4,000m. **ID** Similar to other ground-tyrants. Small, delicate-looking bird with no obvious markings, usually with fluffy plumage and washed-out brown crown. **Habitat** Altiplano. At wetlands and nearby. **Voice** A short, high and quiet contact call, audible only when very close, repeated every three seconds (*fueet*), which can become aggressive during agonistic interactions. **Where to see** Las Cuevas and Parinacota bofedales, and Chungará lake (Arica y Parinacota Region).

PLATE 74: SHRIKE-TYRANTS AND ALLIES

SHRIKE-TYRANTS Large tyrants, usually with relatively large hooked bills. Aggressive, and their diet includes small vertebrates (lizards, mice). Very similar, and identification often relies on overall structure, tail colour and bill shape.

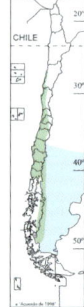

Great Shrike-Tyrant
Agriornis lividus   R
L 26–30cm. Sexes identical. **C** 0–2,000m. **A** 0–700m. **ID** Similar to other shrike-tyrants. Large-headed, with a heavy bill and horn-coloured lower mandible, heavily streaked throat, rufous undertail-coverts and dark tail. **Habitat** Gorges and semi-arid coastal plateaus. Also edges of forest, urban areas and Andean foothills. **Voice** A short, high and dry *tchwek... tchwek...* given at rather long intervals, although it is sometimes given as a quick *tchwek-tchwek-tchwek*.... **Where to see** Conchuca lagoon (Coquimbo Region) Zapallar and Mantagua (Valparaíso Region), Torres del Paine National Park (Magallanes Region).

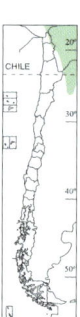

White-tailed Shrike-Tyrant
Agriornis albicauda   R
L 25–27cm. Sexes identical. **N** 4,000–4,600m. **ID** Similar to other shrike-tyrants. Large-headed, with a heavy bill and horn-coloured lower mandible, but throat streaking weakly defined and has white tail except for two central feathers. **Habitat** Altiplano. Rocky slopes near wetlands. **Voice** A short monosyllabic *pewp* heard in early morning. Also an even whistle of 2–3 syllables, repeated at well-spaced intervals (*few-few-feww* ...), which can become an accelerated *fewt-teet-teet* .... **Where to see** Parinacota (Arica y Parinacota Region). Very uncommon.

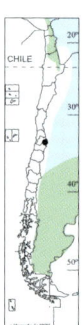

Grey-bellied Shrike-Tyrant
Agriornis micropterus   R
L 23–25cm. Sexes identical. **N** 0–4,200m. **A** 100–500m. **ID** Similar to other shrike-tyrants. Long-bodied. Horn-coloured lower mandible and long, dark tail. Throat streaking heavier and better defined in southern race. **Habitat** Arid and semi-arid shrub-steppe, and adjacent rocks; even urban areas. **Voice** Largely silent and apparently only vocal when nesting. A short, soft *peep* given at irregular intervals. **Where to see** Road from San Pedro to Toconao (Antofagasta Region), Torres del Paine National Park (Magallanes Region).

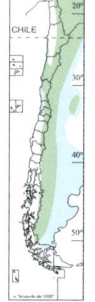

Black-billed Shrike-Tyrant
Agriornis montanus   R
L 23–25cm. Sexes identical. **N** 0–4,600m. **C & S** 0–3,000m. **ID** Similar to other shrike-tyrants. White tail with dark base (varies racially) and dark central feathers. Slender black bill. Pale brownish-white iris. **Habitat** Semi-arid areas with bushy cover, rocky slopes and high-Andean or Patagonian steppe. Valleys and arid coast in central and northern Chile. **Voice** Best known is a loud, strident, human-like whistle of two syllables that lasts less than 1 second and is repeated every 10–15 seconds. The first syllable is ascending and the other descending tone, although sometimes the difference is imperceptible (*fuuuiit-teeuu...*). Sings very early in the morning. **Where to see** Parinacota (Arica y Parinacota Region), Choros gorge (Coquimbo Region), Vallenar (Atacama Region), Farellones (Metropolitan Region); Torres del Paine National Park (Magallanes Region).

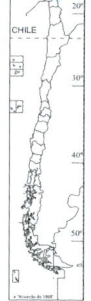

Rufous-webbed Bush-Tyrant
Polioxolmis rufipennis   R
L 18–21cm. Sexes identical. 3,100–4,400m. **ID** Unmistakable. Recalls a shrike-tyrant, but has cinnamon wings and tail, visible in flight. **Habitat** Associated with Queñoa forest (*Polylepis* spp.) and *Puya* spp. In rocky areas. **Voice** A short, high, quiet and irregularly repeated *pweeuu*. **Where to see** Very rare. Altos de Pachama Queñoa forest (Arica y Parinacota Region), Mucomucone (Tarapacá Region). **Note** Breeding records in 1998 and 1999 confirm resident status.

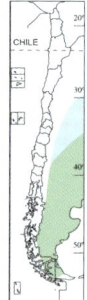

Chocolate-vented Tyrant
Neoxolmis rufiventris   R
L 22–24cm. Sexes identical. 0–500m. **ID** Unmistakable. Flight very different from other similar-sized passerines, being almost shorebird-like. **Habitat** Patagonian steppe, grassland with cushion plants, low scrubby areas. **Voice** Two main calls. One, given perched or in flight, possibly to maintain contact, is quiet and given up to three times (*tchip... tchip-tchip... tchip-tchipchip...*), without any obvious sequence. The song is a high, penetrating disyllabic whistle, which can become more complex; the first syllable ascends and the second is slightly descending (*feeuu-fuiii...*). **Where to see** Sierra Baguales and Pali Ayke National Park (Magallanes Region).

158

PLATE 75: CHAT-TYRANTS, FLYCATCHERS, KISKADEE AND KINGBIRD

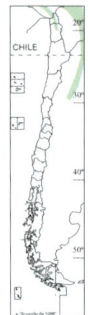

D'Orbigny's Chat-Tyrant
Ochthoeca oenanthoides R
L 15–17cm. Sexes identical. 3,000–4,300m. **ID** Upperparts similar to White-browed Chat-Tyrant. Intense rufous belly. **Habitat** Gorges with scrub, Queñoa forest (*Polylepis* spp.), and rocky slopes. Enters high-Andean hamlets. **Voice** During early morning, singly or in pairs gives a somewhat complex, long, and aggressive-sounding vocalisation that starts *tew-wit*, repeated after c.2 seconds, and the whole continues for several seconds (*tew-wit … tew-wit… tew-wu-wit - tew-wu-wit – tew-wu-wit…*). Very soft calls, probably in contact (*pip*). **Where to see** Altos de Pachama Queñoa forest, Guallatire (Arica y Parinacota Region).

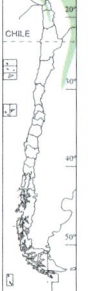

White-browed Chat-Tyrant
Ochthoeca leucophrys R
L 14–16cm. Sexes identical. 2,000–3,600m. **ID** Upperparts similar to D'Orbigny's Chat-Tyrant. Underparts pale grey or off-white. **Habitat** Streams with scrubby sides. **Voice** A high monosyllable, sometimes very soft, in contact call, or aggressively repeated (*p'tew*). **Where to see** Socoroma, Putre and Chapiquiña (Arica y Parinacota Region), Mamiña (Tarapacá Region).

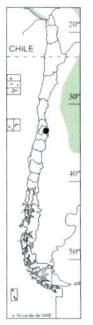

Rufescent Flycatcher
Myiophobus rufescens R
L 11–13cm. Sexes similar. 0–300m. **ID** Unmistakable. A small bird that often skulks in bushes. Unmarked, all-rufous underparts, with two rufous wingbars. **Habitat** Scrubby environments, in semi-arid and semi-desert areas. **Voice** Territorial song complex, sometimes given by pair. One bird gives the same metallic note over and over (*KLI-KLI … Kli-kli … Kli-kli …*), while the other produces a very soft, long trill (*trrrrrrrrrrrrrrrrrr….*). **Where to see** Lluta and Azapa Valleys, Camarones canyon (Arica y Parinacota Region). **Note** Formerly treated as conspecific with Bran-coloured Flycatcher *M. fasciatus*.

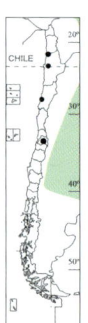

Tropical Kingbird
Tyrannus melancholicus V
L 20–22cm. Sexes identical. 0–3,100m. **ID** Unmistakable in Chile. A large kingbird with mainly yellowish underparts, a grey head and largely olive underparts. **Habitat** Most environments. **Voice** Very vocal. Song, given before sunrise, comprises several high, slightly ascending syllables (*twetitit… tweetit-twetwet…*). During the day gives a sharp, wheezing *siriri… siriri... Siriri…*), repeated several times. **Where to see** No specific sites. Mostly observed between Arica and Antofagasta.

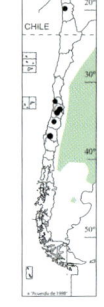

Great Kiskadee
Pitangus sulphuratus V
L 22–25cm. Sexes identical. 0–3,300m. **ID** Unmistakable in Chile. Large tyrant-flycatcher with a big bill, rufous wings and tail, yellow underparts and boldly marked head pattern. **Habitat** Great diversity of environments, near water. Even in cities. **Voice** Outside Chile, most frequent call is a high shriek in contact, mostly given early morning, sometimes forming a short, well-spaced series (*weiuuuu …*), and a loud, high-pitched *pe-tiew-fieew*, repeated partially or in its entirety, sometimes singly. **Where to see** Accidental. Usually in the north.

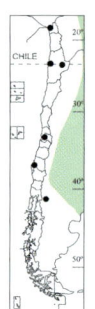

Fork-tailed Flycatcher
Tyrannus savana V
L 38–40cm. Sexes similar. 0–4,250m. **ID** Similar to Eastern Kingbird. Very long tail and black cap to eye level, with partial white collar. **Habitat** Open areas, but at edges of woodland, usually in areas with only scattered trees. **Voice** In contact gives a metallic click at intervals (*tchip… chip…*); song is a series of similar clicks that accelerate and become more complex. **Where to see** Accidental. Usually in the north.

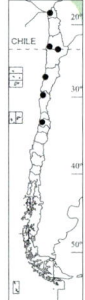

Eastern Kingbird
Tyrannus tyrannus V
L 19–21cm. Sexes similar. 0–1.800m. **ID** Similar to Fork-tailed Flycatcher. Short black tail with a white tip, and blackish head to eye level which reaches the neck. **Habitat** Most environments, but seems to prefer wooded and semi-wooded areas. Even in cities. **Voice** Silent in South America (breeds in North America). **Where to see** No specific sites. More likely between Arica and Antofagasta.

PLATE 76: FLYCATCHER, ELAENIA, TYRANT AND DIUCON

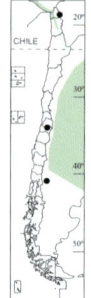

Vermilion Flycatcher
Pyrocephalus rubinus   R
L 13–15cm. Sexually dimorphic. **N** 0–500m.
ID Unmistakable. Juvenile and adult female similar, but former has yellowish belly. **Habitat** Open areas, with scrub or low trees. Valleys with water and agriculture. **Voice** A rather quiet and irregularly repeated *peek* in contact. Also a high-pitched song that lasts less than 3 seconds and is repeated every 2–3 seconds, and which becomes a slight trill in finale (*plt-pit-twirir*). **Where to see** Lluta, Azapa and Chaca Valleys (Arica y Parinacota Region). [Alt. Common Vermilion Flycatcher]

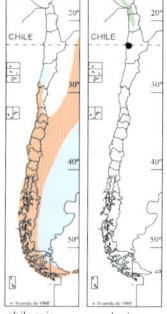

White-crested Elaenia
Elaenia albiceps   R
L 14–15cm. Sexes identical. **N** 0–2,000m.
C 0–2,600m. **S** & **A** 0–1,300m.
ID Unmistakable. White erectile tuft on crown. **Habitat** Arboreal. Northern race occurs in Peruvian Peppertree (*Schinus molle*) woodland and plantations. In central and southern Chile in sclerophyll, temperate or wet forest, gorges with dense scrub and parks in the cities.
Voice Frequently calls, and is a typical sound of Chilean temperate forests: a single-note, melancholic *feeww*, given

chilensis   *modesta*   every 5–7 seconds, all day long. Song,
however, is usually heard only in early morning; two dry notes (*few-few*). **Where to see** Year-round in the north. Azapa Archaeological Museum, Lluta and Chaca Valleys (Arica y Parinacota Region). In central and southern Chile only September–March. Metropolitan Park (Metropolitan Region), La Campana National Park (Valparaíso Region), Termas de Chillán (Ñuble Region), Nahuelbuta National Park (Araucanía Region), Chiloé Island (Los Lagos Region). **Note** Race *modesta* is sometimes considered to be a separate species.

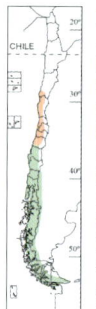

Patagonian Tyrant
Colorhamphus parvirostris   R
L 12–14cm 0–2,000m. Sexes identical.
ID Unmistakable. Small, grey-headed bird with blackish ear-coverts and two rufous wingbars.
Habitat Southern temperate forest. Moves north in winter to sclerophyll forests in the Andean foothills and coastal range (especially *Quillaja* sp., *Maytenus* sp.). **Voice** Frequently heard during breeding season is a mournful whistle, which varies in length, repeated every 3–4 seconds (*fiiiiiiii...*) and sometimes interspersed with a short, dry and accented *fi-few*. **Where to see** In winter, Campana National Park and coastal forests (Valparaíso Region). In spring/summer, Nahuelbuta National Park (Araucanía Region), Huerquehue National Park (Los Lagos Region), etc.

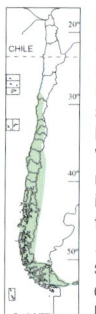

Fire-eyed Diucon
Xolmis pyrope   R
L 19–21cm. Sexes identical. 0–2,000m.
ID Unmistakable. Large head and red eyes. Juvenile similar but has dark iris and slightly grey-streaked breast. **Habitat** Forest, although prefers clearings. Will enter cities. **Voice** Territorial song given in early morning, rarely later in the day, comprises a soft introductory syllable, followed c.1 second later by three rapid notes, the last high and accented (*fwet... fu-Fi-Tew...*). In interactions with conspecifics gives soft monosyllables (*fweet*). **Where to see** In winter, coastal areas (Coquimbo Region, Valpariso Region, Metropolitan Region), La Campana National Park (Valparaíso Region), Metropolitan Park (Metropolitan Region). In summer, Nahuelbuta National Park (Araucanía Region), Chiloé Island (Los Lagos Region).

PLATE 77: THRUSHES AND MOCKINGBIRDS

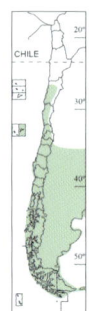

Austral Thrush
Turdus falcklandii R
L 24–27cm. Sexes similar. 0–2,600m. **ID** Recalls Great Shrike-Tyrant and Creamy-bellied Thrush. Yellow-orange bill and legs, and streaked white throat. **Habitat** Various environments, including cities, but requires shady and humid vegetation. **Voice** Apparently in contact, gives a sharp, penetrating and somewhat scratchy *tweeck* every 2–3 seconds, becoming louder in alarm. Also in alarm, a hoarse repeated *trrrweet..trrrweet... trrrweet...*. In spring and summer, sings at dawn, even well before sunrise, and at dusk; similar to other *Turdus* spp., with whistles and trills forming an almost strident but pleasant melody. **Where to see** Squares and parks in cities, damp ravines in central Chile. Common throughout range.

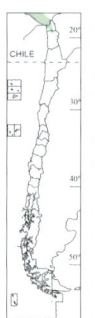

Chiguanco Thrush
Turdus chiguanco R
L 26–29cm. Sexes similar. 2,500–3,700m. **ID** Similar to allopatric Sombre Thrush. Lacks eye-ring. **Habitat** Ravines with scrub and trees, cultivated areas and open areas around water. Villages. **Voice** Most frequently heard is a short, powerful *pweu... pweu...* given repeatedly, and sometimes becomes a more constant but 'reluctant' *pweu-pweu-pweu-pweu-pweu-pweu...*), or a higher and more drawn-out *peewww- peewww - peewww - peewww - peewww...* in alarm. Song, at dusk and dawn, comprises a typical thrush series of whistles and trills, usually given by male from a high perch. **Where to see** Mountains of the Arica y Parinacota Region, e.g. Socoroma, Putre, Chapiquiña, etc.

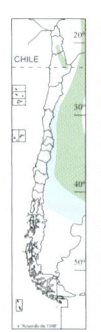

Sombre Thrush
Turdus anthracinus R
L 26–29cm. Sexes similar. 2,000–3,000m. **ID** Similar to allopatric Chiguanco Thrush. Note orange eye-ring and very dark body colour. **Habitat** Scrub and trees, agricultural areas and open ground near water. Villages. Arboreal. **Voice** A short, sharp, monosyllabic *pew-pew* given repeatedly. Song is individually variable, and typical of all Chilean thrushes. **Where to see** San Pedro de Atacama, Quitor, Toconao and Jere gorge (Antofagasta Region). **Note** Formerly treated as conspecific with Chiguanco Thrush *T. chiguanco*.

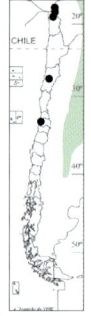

Creamy-bellied Thrush
Turdus amaurochalinus V
L 22–24cm. Sexes similar. 0–3,600m. **ID** Similar to Great Shrike-Tyrant and Austral Thrush. Generally brown, with yellowish bill and dark legs. **Habitat** Scrub and open forest. Very arboreal. **Voice** Most frequently heard is a low-pitched whistle in contact, sometimes doubled (*whip... whip-whip ...*) and becomes faster in alarm. Also in contact, a powerful, cat-like *peow... meow*. Very elaborate and pleasing song comprises whistles, trills and twitters typical of all *Turdus* thrushes. **Where to see** Accidental.

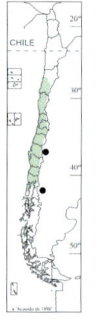

Chilean Mockingbird
Mimus thenca R
L 27–29cm. Sexes identical. 0–2,400m. **ID** Recalls other mockingbirds. Brownish-grey, with well-marked malar and no white in primaries or secondaries. **Habitat** Scrubby vegetation and scattered trees, forest edges, semi-arid areas and mountain slopes. **Voice** Sings frequently when breeding. A long, varied and musical song, with repeated stanzas, and alternates harsh syllables with high-pitched whistles and very melodic trills. One of the most characteristic songs of semi-arid areas in central Chile. Also gives hoarse, raspy notes in aggressive interactions (*keeeck-keeeck-keeeck...*). **Where to see** Abundant, even in cities. Coastal range (Coquimbo Region–O'Higgins Region), road to Farellones and Maipo canyon (Metropolitan Region), road to Los Cipreses National Reserve (O'Higgins Region).

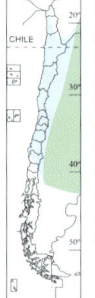

White-banded Mockingbird
Mimus triurus V
L 22–24cm. Sexes identical. 0–2,600m. **ID** Recalls other mockingbirds. No malar, and has white outer tail feathers, as well as some secondaries and wing-coverts, both very evident in flight. **Habitat** Various environments. Shrubby areas, thickets and steppe. **Voice** Song like that of other mockingbirds, and prolonged. In alarm or contact, gives a *tcheit ... tcheit*, repeated every 3–4 seconds. **Where to see** Accidental.

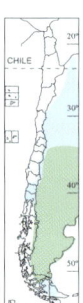

Patagonian Mockingbird
Mimus patagonicus R
L 22–24cm. Sexes identical. 0–1,800m. **ID** Recalls other mockingbirds. Lacks malar and has no white in primaries or secondaries. **Habitat** Patagonian thickets and steppe. **Voice** Calls frequently. Song recalls that of other mockingbirds. Like Chilean Mockingbird, but always higher and imitates other species. In alarm gives a *pew* note, repeated sporadically every 5–10 seconds. In aggression, gives a scratchy or grunting *cgrrrrr-cgrrrrr-cgrrrrr-cgrrrrr -cgrrrrrcgrrrrr-cgrrrrr...*, comprising 7–8 syllables, and repeated every c.5 seconds. **Where to see** Sierra Baguales and Torres del Paine National Park (Magallanes Region).

164

PLATE 78: PLANTCUTTER, MEADOWLARKS AND BLACKBIRD

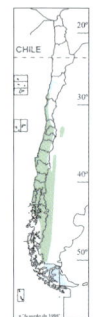

Rufous-tailed Plantcutter
Phytotoma rara   **R**
L 18–20cm. Sexually dimorphic. 0–2,600m. **ID** Unmistakable. Both males and females have peculiar shape on head and bill. Adult male has brick-red underparts, which are unique and diagnostic of this species. Adult female is brown, heavily streaked below. Both sexes and all ages have red eyes. Juvenile like female, but with darker eyes. **Habitat** Forest, sclerophyll scrub, crops. Even in cities. **Voice** Frequently heard during breeding season. Given from a high perch, the very unusual song consists of a slightly accelerating introduction, which varies in length, 3–4 to 17 repetitions (*Tck...*), followed by a mechanical-sounding rattle, the whole rather hoarse and metallic, and sometimes part of the introduction is repeated at the end (*Kaekkaek-kaek-kaek-kaek-krrrrrrrrrrrr... kaek-kaek-kaek...*). **Where to see** Rio Maipo wetland (Valparaíso Region), Viña del Mar (Valparaíso Region), Metropolitan Park (Metropolitan Region), Mocha Island (Bio-Bio Region), Chiloé Island (Los Lagos Region).

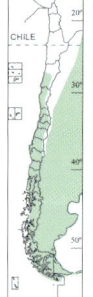

Long-tailed Meadowlark
Leistes loyca   **R**
L 24–28cm. Sexually dimorphic. 0–2,500m. **ID** Unmistakable in range. Adult male has vivid red breast and long white supercilium. Adult female and juvenile are considerably duller, similar to Yellow-winged Blackbird, but with a pink or reddish tinge to underparts and a long sharp bill. **Habitat** Open areas, grassland and forest edge. Even in cities. **Voice** Mostly heard is a monosyllabic, short dry croak (*kiu*). Song varies, but is always loud and given by male from a slightly elevated perch, and sometimes joined by female in duet. Has three parts, the first comprising three whistles, followed by three shorter, strident whistles, and a scratchy ending (*feet-feet-feet — few-few-few — tweeeeee...*). **Where to see** Road to Farellones, road to El Yeso Valley (Metropolitan Region), coastal range (Coquimbo Region–Maule Region), Torres del Paine National Park (Magallanes Region).

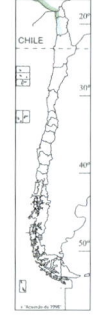

Peruvian Meadowlark
Leistes bellicosus   **R**
L 20–22cm. Sexually dimorphic. 0–2,500m. **ID** Unmistakable in range. The northern counterpart of Long-tailed Meadowlark, with very similar plumage and levels of sexual dimorphism. **Habitat** Semi-arid areas with scrubby vegetation. Open areas and grassland. Harvested fields. **Voice** When breeding, male sings from a high perch; a whistle, followed by a peep and ending in a hoarse trill (*feeee-tweutweutweu-tugheeeeii...*), lasting c.2 seconds and repeated at variable intervals, usually 8–10 seconds. **Where to see** Lluta, Azapa and Chaca Valleys (Arica y Parinacota Region).

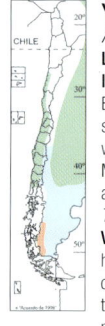

Yellow-winged Blackbird
Agelasticus thilius   **R**
L 17–20cm. Sexually dimorphic. 0–2,000m. **ID** Male is similar to male Shiny Cowbird and Austral Blackbird. No bluish gloss and has yellow spot on smaller wing-coverts, not always visible in flight or when perched. Female recalls female Long-tailed Meadowlark, but is all brown with heavy streaks and an obvious pale eyebrow. **Habitat** Reeds (*Scirpus* sp., *Typha* sp.), but will enter crop fields and grassland. **Voice** Male frequently calls when breeding, perched high and vibrating the wings, highlighting the yellow coverts. Song starts with several short notes, then two prolonged and intense syllables with a slightly nasal quality (*peep ... preep ... trreee-leeeee...*). Also a short twitter and trills. Another little-known song is more elaborate and recalls other icterids, given early morning or at dusk, by groups. **Where to see** El Peral lagoon and Rio Maipo wetland (Valparaíso Region), Lampa/Batuco system (Metropolitan Region). Any wetland with reeds.

PLATE 79: COWBIRDS, BLACKBIRD AND ANI

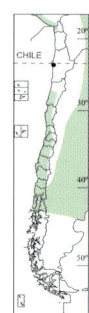

Shiny Cowbird
Molothrus bonariensis   R

L 19–22cm. Sexually dimorphic. 0–2,000m. **ID** Male is similar to Screaming Cowbird, Austral Blackbird and male Yellow-winged Blackbird. A blackbird with obvious bluish to greenish, metallic gloss, especially on wings. **Habitat** Open areas, grassland and scattered trees. **Voice** Rather quiet, but has a very elaborate and complex song, comprising a mix of trills, twitters and whistles forming a harmonious and attractive whole. **Where to see** In winter in flocks in cities such as Viña del Mar (Valparaíso Region), Santiago (Metropolitan Region) and Chillán (Ñuble Region). In spring/summer widespread. Range expanding.

Screaming Cowbird
Molothrus rufoaxillaris   R

L 19–22cm. Sexes identical. 0–2,000m. **ID** Similar to male Shiny Cowbird, Austral Blackbird and male Yellow-winged Blackbird. Bluish-black, with a short conical bill, and a rufous stain on the underwing-coverts, visible only in flight. Juvenile has a brown body and head, and tan-coloured wings. **Habitat** Farms, pastures, scrub and open forests. **Voice** Song rather disjointed and lacks a continuous melody, combining trills, twitters and monosyllables, sometimes recalling flowing waters, often ending each part with a high note, and many times initiated with a trill that becomes an extremely high note. Contact calls short, also sometimes starting as a trill and abruptly becoming very high-pitched. **Where to see** San Fernando, Agua Buena (O'Higgins Region). Possibly expanding.

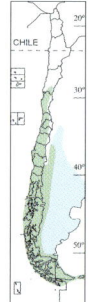

Austral Blackbird
Curaeus curaeus   R

L 26–30cm. Sexes identical. 0–2,000m. **ID** Similar to male Shiny Cowbird and male Yellow-winged Blackbird. A dull blackbird, without blue gloss, commonly seen in noisy flocks. **Habitat** Various environments. Open areas with scrub and very dry vegetation to temperate southern forests and city parks. **Voice** Very melodic and complex song, given frequently when breeding, combines soft notes, whistles and squawks; can recall the sound of water flowing over stones. Highly characteristic contact call is metallic and powerful, usually given by flocks (*peow…peow…tok-ahk-tchak* …). **Where to see** Road to Farellones, Mahuida Park, and El Yeso Valley (Metropolitan Region). Camping areas in Los Cipreses National Reserve (O'Higgins Region), Nahuelbuta National Park (Araucanía Region), Chiloé National Park (Los Lagos Region), Torres del Paine National Park (Magallanes Region).

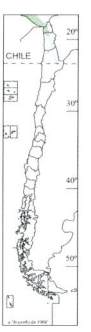

Groove-billed Ani
Crotophaga sulcirostris   R

L 30–32cm. Sexes identical. 0–1,000m. **ID** Unmistakable. Long-bodied, long-tailed, all-black bird, with chunky bill and large head but flat crown. The only bird with that shape, colour and behaviour in its range. Rarely seen alone; always in pairs or flocks. **Habitat** Various environments near water and with cattle. Tree-lined areas, riverbanks, lakes and lagoons, farmland and pasture, and semi-arid areas. **Voice** Two-syllable call, melodious and high, and sometimes drawn-out (*peet-tuiiiiii* ...), or repeated (*p't-tui … p't-tui … p't-tui … p't-tui* ...). In alarm whole flock flies to a nearby tall tree, before returning to their routine. **Where to see** Lluta, Azapa and Camarones Valleys (Arica y Parinacota Region).

PLATE 80: PIPITS AND WRENS

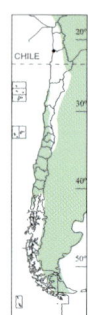

Correndera Pipit
Anthus correndera   **R**
L 14–16cm. Sexes identical. **N** 2,500–4,000m. **C** 0–2,600m. **S & A** 0–500m. **ID** Very similar to Hellmayr's and Yellowish Pipits. Two obvious ochre lines on mantle, a well-marked malar and dense breast streaking. **Habitat** Open grassland and flooded areas near rivers, estuaries and other waterbodies. **Voice** Most often heard is song in display flight, which is high-pitched and repeated c.2–3 times (*twet - tweitweit - twetweiteit - tweteit* ...), and in descent changes to a syllable repeated three times followed by a long trill (*Twetweitweit - trrrrriiiiiiiiii... Twetweitweit - trrrrriiiiiiiii...*), until reaching the ground. Also an irregularly repeated monosyllable, *trrweet*, from the ground, apparently in alarm. **Where to see** Inland lagoons on Atacama Salt Flat (Antofagasta Region), Rio Maipo wetland (Valparaíso Region), Lampa/Batuco system (Metropolitan Region), Torres del Paine National Park and Los Cisnes lake (Magallanes Region).

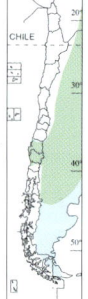

Hellmayr's Pipit
Anthus hellmayri   **R**
L 14–15cm. Sexes identical. **S** 0–1,200m. **ID** Similar to Correndera Pipit. No malar and breast streaking finer and sparser. **Habitat** Open grassland, not on steep slopes. In stony and semi-arid environments. **Voice** Three-syllable call, repeated several times from the ground, usually from cover, the first one slightly trilled (*Trreeet - Teew - Teew*), apparently in contact. **Where to see** Near Temuco, Victoria and Boca Budi (Los Lagos Region).

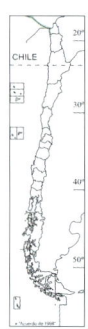

Yellowish Pipit
Anthus lutescens   **V**
L 12–13cm. Sexes identical. 0–100m. **ID** Similar to Correndera Pipit. Brown mantle with four or five diffuse yellowish-ochre lines, a slight malar, densely streaked breast, and relatively long legs. **Habitat** Open grassland, near rivers, estuaries and other waterbodies. **Voice** Song, given in aerial display, starts with a repeated high-pitched note during ascent (*tsiptsip-tsip-tsip-tsip-tsip-tsip* ...), but upon descending changes to a complex trill (*tiglglglglglglglglglglgl...*). **Where to see** Records at Lluta wetland and Camarones gorge (Arica y Parinacota Region).

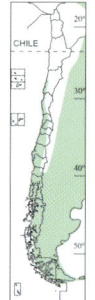

Grass Wren
Cistothorus platensis   **R**
L 10–12cm. Sexes identical. 0–1,500m. **ID** Similar to Wren-like Rushbird and House Wren. Inconspicuous, with brown and ochre striped upperparts (no grey) and a barred tail, held cocked, as well as a narrow supercilium. **Habitat** Waterlogged, humid grassland and damp scrub; around wetlands. **Voice** Contact call recalls wood being knocked together at intervals (*taek ... taek ... taek ...*), and can become louder and intense hum, including an accelerating, high trill (*taek-taek-taek-taek...*). When breeding, song complex and individually variable: several parts, each preceded by a short and variable trill (*TzzTrrrrrrr...*), a somewhat looser trill (*TzzTrTrTrTr...*) and a metallic twitter (*TzzPeePeePeePeePeew...*), with intervals of c.2 seconds between parts, and a variable sequence. **Where to see** Lampa/Batuco system (Metropolitan Region), Rio Maipo wetland (Valparaíso Region), Mocha Island (Bio-Bio Region). **Note** A recent study determined that *C. platensis* included at least eight subspecies groups, which may be elevated to species in future. The form in Chile would be Austral Wren *C. hornensis*.

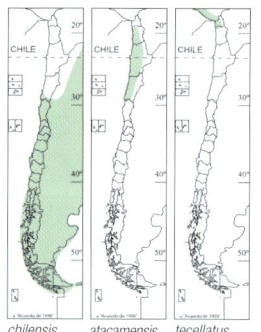

chilensis   *atacamensis*   *tecellatus*

House Wren
Troglodytes aedon   **R**
L 10–13cm. Sexes identical. **N** 0–3,600m. **C** 0–2,500m. **S & A** 0–1,200m. **ID** Similar to Grass Wren. Upperparts uniform brownish or greyish-brown, sometimes very slightly barred, but unstreaked. **Habitat** Wide variety of habitats. Scrub and forest edge, in humid, arid and semi-arid environments. **Voice** Frequently heard. Year-round, in either contact or alarm, gives a two-note *Tchrrrk-Tchrrrk ... Tchrrrk-Tchrrrk*. Also a longer, dry *tchrk-tchrk-tchrk-tchrk-tchrk ... tchrrk ... tchrrk*, like two hard surfaces being scraped together, which starts faster but slows towards the end. When breeding, male has a complex and quite melodic song, given from an obvious perch while vibrating the wings. **Where to see** Very common throughout. Ssp. *tecellatus* in Lluta and Azapa Valleys (Arica y Parinacota Region).

PLATE 81: NEW WORLD WARBLER AND TANAGERS

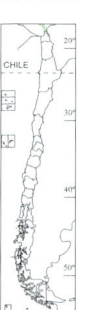

American Redstart
Setophaga ruticilla   V
L 12–13cm. Sexually dimorphic. 0–3,700m. **ID** Unmistakable. Yellow tail with black terminal band and central feathers. **Habitat** In low valleys and pre-puna zone. Sparse woodland. **Voice** Apparently silent in Chile. Breeds in North America. **Where to see** Recorded at Putre (Arica y Parinacota Region) and Baquedano (Antofagasta Region).

Giant Conebill
Conirostrum binghami   R
L 14–17cm. Sexes identical. 3,000–4,600m. **ID** Unmistakable. Unique in size and colour. Sharply pointed bill, blue-grey upperparts, rufous underparts and largely white face are, in combination, wholly distinctive, especially coupled with specialist habitat requirements (*Polylepis* sp. forest). When feeding in *Polylepis* forest, it produces a characteristic sound when chopping and stirring the bark. **Habitat** Exclusively in Queñoa forest (*Polylepis* sp.). **Voice** Gives very soft and high-pitched calls, probably in contact, repeated at intervals while foraging (*tzip - tzip - tzip - tzip - tzip…*). Very elaborate and musical song is long and varied. **Where to see** The Arica y Parinacota Region, at Altos de Pachama, Japu and other Queñoa forests. **Note** Formerly listed as *Oreomanes fraseri*.

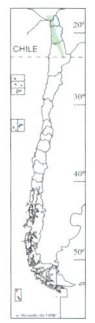

Tamarugo Conebill
Conirostrum tamarugense   R
L 11–13cm. Sexes identical. 0–3,700m. **ID** Similar to Cinereous Conebill. Adult lead-grey, with variable red or rufous marks on face, throat and undertail-coverts. Juvenile mainly pale brown, with streaked breast and sometimes small flecks of red on throat. **Habitat** Mostly in Tamarugo (*Prosopis tamarugo*) woodland, even isolated trees of this species. In non-breeding season gorges and creeks with bushes and scrub, usually in highlands. **Voice** Calls frequently in breeding season, otherwise rather quiet. Pairs or families foraging for food give very high contact calls reminiscent of Cinereous Conebill, but more nasal (*psee-psi-psisisisiiii …..psee-psi-psi-psisisisiiii…*). Also used in aggression. **Where to see** In autumn/winter in Lluta and Azapa Valleys and pre-puna at Putre, Belén and Codpa (Arica y Parinacota Region). In the breeding season, at Pampa del Tamarugal National Reserve (Tarapacá Region), road from San Pedro de Atacama to Toconao (Antofagasta Region).

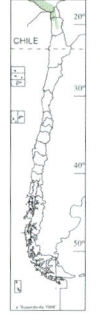

Cinereous Conebill
Conirostrum cinereum   R
L 11–13cm. Sexes identical. 0–3,700m. **ID** Concealed in vegetation, could be confused with Tamarugo Conebill, but note their different calls. Adult has blue-grey back, clear white eyebrow and no red or rufous in plumage. Juvenile mainly ochre, also without rufous. **Habitat** Thickets and wooded areas, even parks and gardens in cities. **Voice** Calls frequently in breeding season. Pairs or families when foraging maintain contact using 7–12 short, nervous, high-pitched notes (*pzeet-pzeet-pzeet-pzeet-pzeet…*), repeated constantly. Occasionally a single, high *pzeit*, repeated several times. Song is high-pitched and aggressive-sounding, usually given from an exposed perch. **Where to see** Arica, Lluta and Azapa Valleys, Putre and Socoroma, in pre-puna zone (Arica y Parinacota Region).

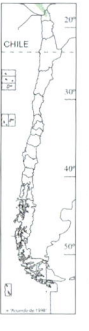

Black-throated Flowerpiercer
Diglossa brunneiventris   R
L 12–14cm. Sexes identical. 3,000–3,600m. **ID** Unmistakable. Small bird with a peculiar and highly-specialised hook-tipped bill (in all ages). Beautiful colours, with mainly black upperparts and throat, rest of underparts deep rufous, and has bluish-grey flanks. **Habitat** Semi-arid and arid areas with shrubs. Scrubby gorges. **Voice** Calls frequently in breeding season, otherwise rather quiet. Song complex, usually given from a high exposed perch, lasts c.1 second and is repeated every 3–4 seconds. **Where to see** Putre, Socoroma, Chapiquiña and elsewhere in pre-puna zone (Arica y Parinacota Region).

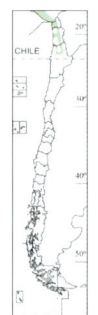

Slender-billed Finch
Xenospingus concolor   R
L 15–17cm. Sexes identical. 0–500m. **ID** Unmistakable. At all ages has very long tail. **Habitat** Semi-arid areas with bushes, shrubs and trees. Usually near water. **Voice** Calls unusual, resembling groans, cracking of trees, etc. An unusually loud *tcheck* in contact, as well as a somewhat growling *grrk … grrk …*, a very loud *pew*, and a complex song comprising whistles, trills, and more peculiar sounds such as those mentioned, in addition to very high notes. **Where to see** Lluta Valley, Azapa Archaeological Museum, Caleta Vitor (Arica y Parinacota Region), Loa River (Antofagasta Region).

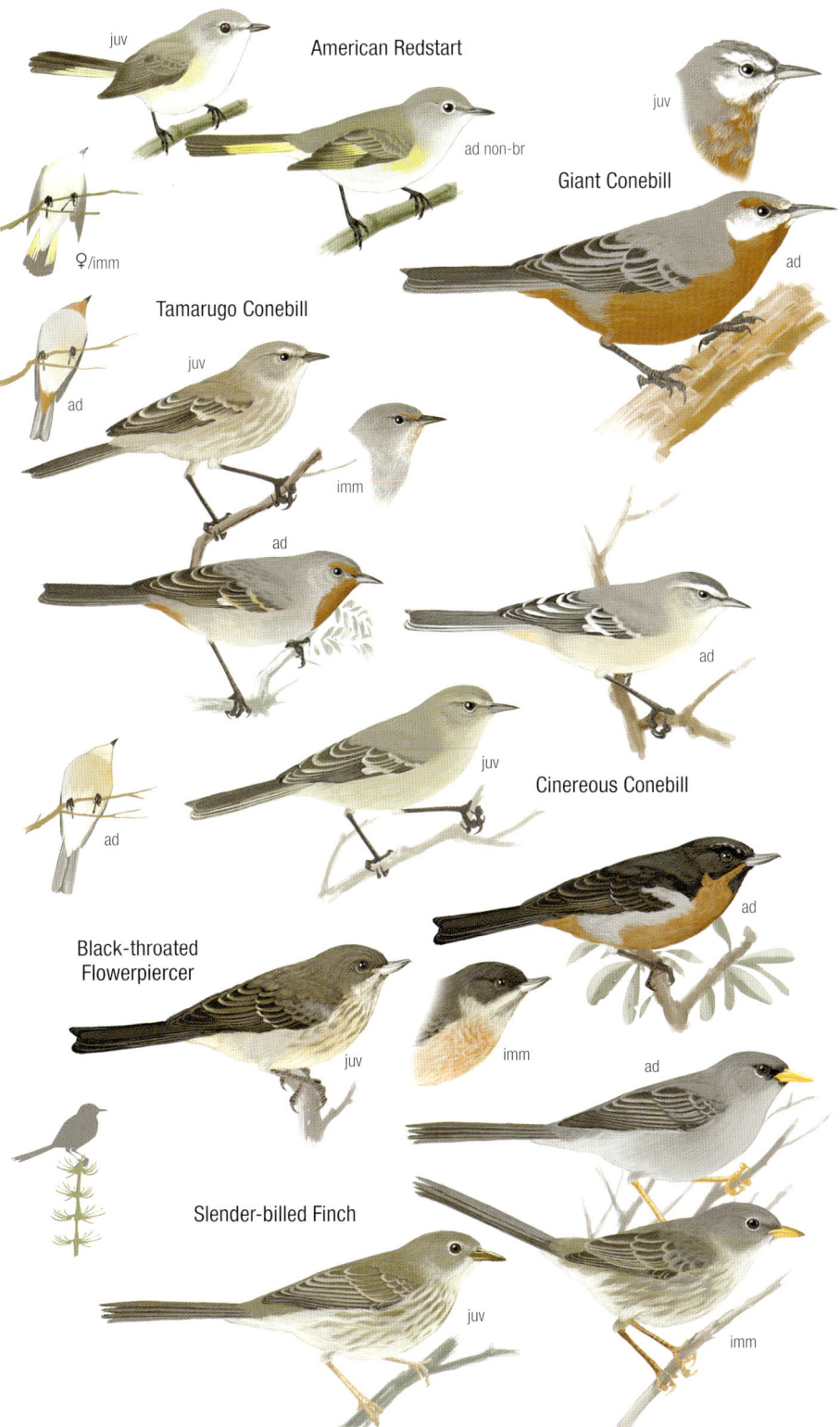

PLATE 82: GRASSQUIT, SEEDEATERS AND TANAGERS

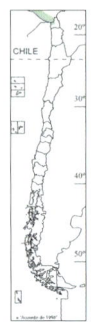

Blue-black Grassquit
Volatinia jacarina R
L 10–12cm. Sexually dimorphic. 0–1,000m.
ID Unmistakable. Adult male entirely glossy blue-black, with white wing-linings obvious in flight, especially display. Female principally dull brown and rather streaked below. Both sexes have somewhat pointed, conical bill. **Habitat** Both arid and humid areas. Garlic crops, onion and corn stubble. **Voice** Song, given by male from highly visible perch, is usually accompanied by a display in which it flutters quickly up c.30cm revealing the white in the wing, before descending to the initial position, and which is repeated many times; a short, metallic, half-buzzing and half-whistle (*Tzzzz-tzzzzz-Feee...*). At close quarters sound made by the wings can also be audible. **Where to see** Lluta and Azapa Valleys (Arica y Parinacota Region).

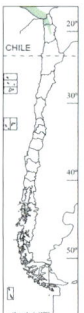

Chestnut-throated Seedeater
Sporophila telasco R
L 10–13cm. Sexually dimorphic. 0–1,000m.
ID Unmistakable. Chunky little seedeater. Male distinctive by virtue of chestnut throat, otherwise largely grey above and white below. Adult female brownish and streaked. **Habitat** Semi-arid and arid environments, with scrub and open areas. **Voice** Male is vocal during breeding season, singing from a prominent perch, six fairly loud syllables, the first three twittering, and the rest hoarse, drawn-out and scratchy, ending in a trill (*tee-tweet-tweetee- teekei-teekei-trrukiki...*). **Where to see** Lluta and Azapa Valleys (Arica y Parinacota Region).

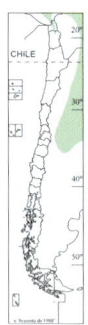

Band-tailed Seedeater
Catamenia analis R
L 11–13cm. Sexually dimorphic. 0–3,700m.
ID Similar to Band-tailed Sierra-Finch. Large bill and rufous undertail-coverts. **Habitat** Semi-arid pre-puna with low vegetation. **Voice** Male gives a long, compressed, high trill, with an almost metallic quality, *pleeleeleeleeleeleelee...* Also a shorter trill, repeated every 2–3 seconds, which is hoarser and quieter in tone, but sounds equally monotonous and compressed (*trrrrrrrrrrreee...*). **Where to see** Socoroma and Putre (Arica y Parinacota Region).

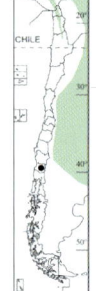

Blue-and-yellow Tanager
Pipraeidea bonariensis R
L 16–18cm. Sexually dimorphic. 2,500–3,700m.
ID Unmistakable. Very easily identified. Male predominantly bluish and yellow-green. Female is much duller, with all colours appearing very washed-out. **Habitat** Scrub with scattered trees, in arid and semi-arid areas. **Voice** Calls frequently during breeding season. The long and relatively complex song, given from an exposed perch, usually commences with a repeated strident *psee-tew - pseetew- pseetew - pseetew...* given 3–4 times, before continuing with other whistles and trills. The same disyllabic note is used in contact, albeit lower and slightly drawn-out (*Pssee-tew - psee-tew...*), but also repeated 3–4 times. **Where to see** Socoroma, Putre (Arica y Parinacota Region), Chuzmisa gorge (Tarapacá Region).

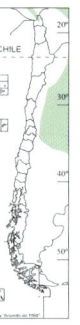

Golden-billed Saltator
Saltator aurantiirostris R
L 20–22cm. Sexes identical. 3,000–3,700m.
ID Unmistakable passerine with chunky orangey bill, pale chin and postocular supercilium, buffy underparts, grey-brown upperparts, and a blackish face and throat. Juvenile retains basic pattern but has much duller bill. **Habitat** Semi-arid pre-puna ravines with thickets, especially Yara (*Dunalia spinosa*). **Voice** Usually remains concealed, even when singing, although it also perches on rocks or tall trees (e.g. ornamental pines and eucalyptus). The song is complex and melodic, and clearly recalls that of Common Diuca-Finch. **Where to see** Socoroma and Putre (Arica y Parinacota Region).

PLATE 83: SIERRA-FINCHES I AND DIUCA-FINCH

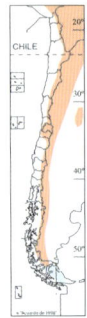

Plumbeous Sierra-Finch
Geospizopsis unicolor R
L 15–18cm. Sexually dimorphic. **N** 3,000-4,600m. **C** 600–3,500m. **S & A** 300–1,500m. **ID** Male unmistakable. Juvenile and female can be confused with juvenile Ash-breasted Sierra-Finch and female Band-tailed Sierra-Finch. Note streaked underparts (to belly) and dark bill, sometimes with slightly horn-coloured lower mandible. **Habitat** Low thickets, grassy and rocky environments. **Voice** Calls little and quietly. Song varies; in the north, a very high, discreet and simple *tet - tweet - tweeiii...*, given over and over from atop a rock or a bush, but more elaborate, musical, longer and complex in central and southern Chile. **Where to see** Las Cuevas, Lauca National Park and Surire (Arica y Parinacota Region), El Yeso Valley (Metropolitan Region), Termas de Chillán (Ñuble Region).

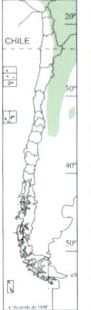

Ash-breasted Sierra-Finch
Geospizopsis plebejus R
L 11–13cm. Sexes similar. **N** 2,500–4,500m. **ID** Similar to Plumbeous Sierra-Finch. Short, sharp bill, white to off-white eyebrow, white throat, compact and rotund body. **Habitat** High-Andean rocky slopes, high-Andean bogs (bofedales), steppes, bushes (Tolar) and Queñoa forests (*Polylepis* spp.). **Voice** Calls little and quietly. Contact call, very similar to other sierra-finches, is repeated irregularly while foraging (*tzip*). Song initially comprises a long, quiet, somewhat buzzing trill, then a disyllabic high chirp, which is louder than the trill, and can be given once or twice (*treeeeeee... tew-tew tew-tew*). **Where to see** Parinacota and Chungará lake (Arica y Parinacota Region), Ollagüe, road to Talabre and Socaire (Antofagasta Region).

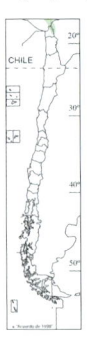

White-throated Sierra-Finch
Idiopsar erythronotus R
L 15–18cm. Sexes identical. **N** 4,000–4,300m. **ID** Adult similar to White-winged Diuca-Finch and juvenile to Red-backed Sierra-Finch. Adult has whitish throat and dark primaries without any white. Juvenile has mantle cinnamon (not reddish-brown or rufous) streaked dark. **Habitat** Highlands. Commonly near water, in high-Andean bogs (bofedales) and nearby areas. **Voice** Usually gives a high-pitched but quiet *peee*. **Where to see** Taipicahue River, Las Cuevas and Parinacota (Arica y Parinacota Region).

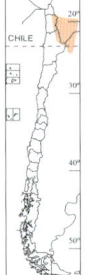

Red-backed Sierra-Finch
Idiopsar dorsalis R
L 15–18cm. Sexes identical. **N** 4,000–5,000m. **ID** Adult similar to juvenile White-throated Sierra-Finch. Warm reddish-brown or rufous mantle without streaks, and slightly rusty flanks. Juveniles of both species very similar, but in present species is much warmer, but equally streaked, and usually has slightly rusty flanks. **Habitat** Restricted to high Andes, near water and high-Andean bogs (bofedales). Also steppe and rocky slopes. **Voice** In contact a high-pitched but quiet monosyllabic *peep ... peep ... peow ...*, which is delivered more rapidly and insistently in alarm; same call is used singly on taking flight. Song apparently undescribed. **Where to see** Tara Salt Flat, Putana River and Tatio hot springs (Antofagasta Region). Generally near bofedales, but abundance varies.

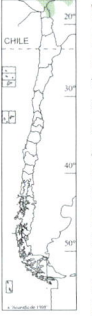

White-winged Diuca-Finch
Idiopsar speculifera R
L 17–19cm. Sexes identical. **N** >4,000m. **ID** Similar to White-throated Sierra-Finch. Clean white throat and white below eyes, contrasting grey body, and a conspicuous white spot on wing. **Habitat** High-Andean bogs (bofedales) and riverbanks. **Voice** Rather quiet. A soft monosyllable, which sometimes recalls a human whistle and can possess an interrogative quality (*tweet* or *pseet*) is used by pair members foraging apart to maintain contact, and is also given in flight. Song more elaborate but still relatively simple, is rarely heard and comprises three parts: first a series of whistles similar to the call but more musical, followed by a c.1-second pause, then another whistle, a second pause, and finally two more whistles, one short and the other compound (*tweet-tweet-tweetweet- twee tweet ... tweet ... .. tweet...tweetweetweet*), and the whole repeated. **Where to see** The Arica y Parinacota Region. Las Cuevas, Parinacota and Guallatire bofedales.

PLATE 84: SPARROWS, DIUCA-FINCH AND SIERRA-FINCHES II

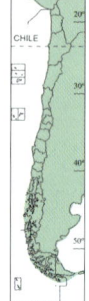

Rufous-collared Sparrow
Zonotrichia capensis   **R**
L 14–16cm. Sexes identical. **N** 0–4,200m. **C** 0–3,000m. **S & A** 0–1,300m. **ID** Grey head (usually with heavy black stripes) and an erectile crest, plus rufous hindcollar. **Habitat** Present throughout Chile and in all habitats. **Voice** Characteristic and well known, despite differences in structure between different populations (and subspecies), with dialects. Song is a high, four-syllable, very melodic *tzeew-wit-tew-tew-tiiiii...*, which can vary slightly, but ssp. *peruviensis*, in northern pre-puna (Putre, Socoroma, etc.) usually has only three notes and a different intonation (*pae-peee-taaa* ...). Juvenile commonly gives a high-pitched buzz, when begging food (*tzeeeeit ... tzeeeeit* ...). Very common for adult to sing at night, with a different voice, but not usually answered by other individuals. **Where to see** Virtually anywhere, including large cities.

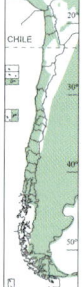

House Sparrow
Passer domesticus   **Ri**
L 14–16cm. Sexually dimorphic. **N** 0–3,600m. **C** 0–2,000m. **S & A** 0–1,000m. **ID** Male similar to Rufous-collared Sparrow, but has black throat and upper breast, plus contrasting white cheeks and throat. Female is mainly brown, with horn-coloured bill and legs. **Habitat** Commensal with human settlements. **Voice** Noisy. Flocks produce a messy chorus. A dry and strong chirp, given persistently from a perch, or while peering from nest (*quik ... quik ... quik ... quik ...*) or a more aggressive *quik ... quik-quik ... quikquu ... quik-quu....* **Where to see** Cities and other settlements.

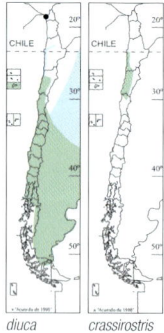

Common Diuca-Finch
Diuca diuca   **R**
L 16–18cm. Sexes similar. 0–2,000m. **ID** Unmistakable. A very boldly but simply patterned finch; mainly dark grey with a large white throat patch and rear underparts cutting high into the centre of the grey breast-band. Adult female averages browner. **Habitat** Thorny thickets, sclerophyll forest, temperate forest edges, agricultural areas, and even cities. **Voice** Calls frequently when breeding. Song is composed of several syllables, usually 3–6, energetic, powerful and apparently independent, but which form a pleasant whole. Males sing early in morning. **Where to see** Fairly common. Pan de Azúcar National Park (Atacama Region), road to Farellones and parks in Santiago (Metropolitan Region), Chiloé Island (Los Lagos Region).

diuca   *crassirostris*

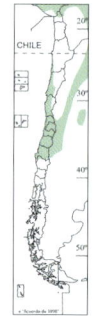

Band-tailed Sierra-Finch
Porphyrospiza alaudina   **R**
L 13–15cm. Sexually dimorphic. **N** 2,500–3,800m. **C** 0–2,800m. **ID** In the north, male could be confused with Band-tailed Seedeater and Mourning Sierra-Finch, but elsewhere only with latter. Male has sharp orange bill and legs, grey throat and breast, unmarked wings and whitish-grey undertail. Female has barely marked malar and ear-coverts concolorous with rest of head. Note tail pattern in flight. **Habitat** Scrub and tall grassland. **Voice** Only regularly vocalises during breeding season. Contact call similar to that of relatives, short and rather weak, and repeated more frequently in alarm (*peep*). Song commences with two short, high monosyllables and ends by repeating a drawn-out disyllabic note three times, the whole being rather musical and lasting c.2 seconds, repeated every 5–6 seconds for long periods, from a perch, in flight or from the ground (*peep - peep - p'teetew -teetew-teetew....*). **Where to see** Road to Nevado Valley (Metropolitan Region), coastal plateaux of central Chile (Coquimbo Region–O'Higgins Region).

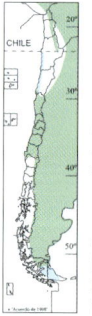

Mourning Sierra-Finch
Phrygilus fruticeti   **R**
L 17–19cm. Sexually dimorphic. **N** 2,000–4,200m. **C** 0–2,800m. **S & A** 0–1,200m. **ID** Male ssp. *coracinus* unmistakable. Ssp. *fruticeti* could be confused with male Band-tailed Sierra-Finch, but note orange bill, black face, throat and breast (variable), two fine white wingbars, and dark tail. Female also has two white wingbars and brown ear-coverts. **Habitat** Sierras and foothills. Areas with bushes and scattered trees. Also Queñoa forests (*Polylepis* sp.) in the north. **Voice** Males constantly vocalise when breeding and are very conspicuous, singing from a high open perch. Song unmistakable, starting as a trill and immediately becoming a metallic buzz, which after less than 1 second becomes a complex trill (*treegnnnnnnn-trweeiwt...*). Contact call recalls a groan, or a cat. **Where to see** Ssp. *coracinus* only near Visviri and Cosapilla (Arica y Parinacota Region); *fruticeti* around Putre (Arica y Parinacota Region), Llanos de Challe National Park (Atacama Region), El Yeso Valley (Metropolitan Region).

PLATE 85: SIERRA-FINCHES III AND BRIDLED FINCHES

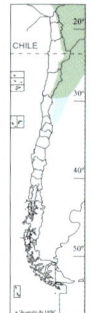

Black-hooded Sierra-Finch
Phrygilus atriceps   R
L 15–17cm. Sexes similar. 2,500–4,600m. **ID** Unmistakable. No other finch with mainly rufous body, and full black hood and wings in Chilean highlands. Adult female averages somewhat duller than male. **Habitat** Andean slopes and high plateaus. Scrubby ravines, low scrub (Tolar) and nearby rocky slopes. Also Queñoa forests (*Polylepis* sp.). **Voice** Song variable but almost always includes a high phrase and *teerweet* ending repeated 3–4 times, sometimes interspersed by other trills and twitters, but can be as simple as *teerweet - teerweet - teerweet - teerweet* …. Song can be repeated several times, every 5–6 seconds. A soft, short and dry *peet … peet* … is given irregularly in contact. **Where to see** Putre, Chungará lake (Arica y Parinacota Region), Machuca (Antofagasta Region).

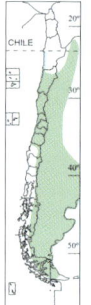

Grey-hooded Sierra-Finch
Phrygilus gayi   R
L 15–17cm. Sexually dimorphic. 0–3,500m. **ID** Male is similar to Patagonian Sierra-Finch. Olive mantle and breast and white undertail-coverts. Female has very strongly marked lateral throat-stripe. **Habitat** Mountains and rocky slopes (ssp. *gayi*), forest edges and areas with scrubby vegetation (ssp. *minor*) and Patagonian steppes (ssp. *caniceps*). **Voice** Contact call given frequently (*peet*), sometimes with a metallic quality, and in alarm is repeated frequently, with a change in tone (*Pit-Pit-Pit* …). Song is a repeated three-syllable phrase (*Peep-peep-Tweet … Peep-peep-Tweet … Peep-peep-Tweet…*). **Where to see** Pan de Azúcar National Park (Atacama Region), Zapallar (Valparaíso Region), El Yeso Valley (Metropolitan Region), Termas de Chillán (Ñuble Region), Torres del Paine National Park (Magallanes Region).

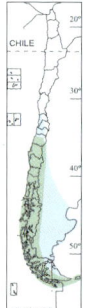

Patagonian Sierra-Finch
Phrygilus patagonicus   R
L 14–16cm. Sexually dimorphic. 0–1,800m. **ID** Male is similar to Grey-hooded Sierra-Finch, but has rufous mantle and more restricted white below. Female has different face and throat pattern. **Habitat** Forests and nearby open areas. Usually in vicinity of water. **Voice** Calls frequently. In contact and alarm (when repeated continuously and accelerates) a variable monosyllable that recalls two stones being hit together (*tick* …) or in alarm, at rate of one note per second (*Tick-Tick-Tick-Tick* …). Can be more clicking (*tchek*). Melodic song always sounds laboured, going up and then down (*Tee-wit-Tee-wit* …), with this phrase repeated 4–10 times, with variations. Song appears to differ between Argentina/austral Chile and central and southern Chile. **Where to see** Nahuelbuta National Park and Cerro Ñielol (Araucanía Region), Chiloé National Park (Los Lagos Region), Torres del Paine National Park (Magallanes Region).

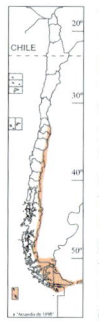

Yellow-bridled Finch
Melanodera xanthogramma   R
L 15–17cm. Sexually dimorphic. **C** 2,800–3,500m. **S** 1,000–1,500m. **A** 0–700m. **ID** At a distance similar to Black-throated Finch. Has grey head with yellowish marks around a black mask and throat, and mostly whitish (not yellow) tail. Juvenile and female mainly greyish-brown with a predominantly whitish tail. See female Patagonian and Grey-hooded Sierra-Finches. **Habitat** Patagonian steppes and foothills, and, in south-central Chile, Andean environments above treeline. **Voice** Generally quiet and vocalisations often inaudible. Gives a fast *fi-fit*, a short whistle (*pseet*) and a longer *psee-psee-psee-psee-pseet*…. Song variable but always somewhat monotonous (*tweet-pli-tcheik … tweet-pli-tcheik …* or *tweet-pli-piw-tcheik … tweet-pli-piw-cheik …*) **Where to see** El Plomo canyon (Metropolitan Region), Ñuble mountains (Ñuble Region), Antillanca volcano and Osorno volcano (Los Lagos Region), Sierra Baguales (Magallanes Region).

White-bridled Finch
Melanodera melanodera   R
L 14–15cm. Sexually dimorphic. 0–500m. **ID** Recalls Yellow-bridled Finch. Male has grey head with obvious white border to black mask and throat, yellow underparts. Wings and tail canary-yellow (very conspicuous). Juvenile and female generally yellowish with a predominantly yellow tail. See female Patagonian and Grey-hooded Sierra-Finches. **Habitat** Patagonian steppe, grassland. **Voice** A very simple but forceful pair of whistles, the first louder and slightly ascending, and the second more muted and somewhat shorter and drier (*Fuit-Fee*), repeated many times. A sequence of whistles, similar to the previous, but of three monosyllables, the first two almost equal, rather energetic, and the third more muted and shorter (*tweee-tweee-pew*…), repeated continuously, and seemingly a more elaborate song. Finally, a longer sequence of short whistles, similar to those already described, interspersing energetic and slightly muted notes. **Where to see** Pali Ayke National Park, steppes near Cerro Sombrero (Magallanes Region). [Alt. Black-throated Finch]

PLATE 86: YELLOW-FINCHES

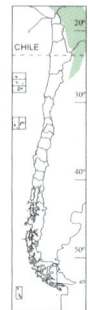

Puna Yellow-Finch
Sicalis lutea R
L 13–14cm. Sexually dimorphic. >4,000m. **ID** Confusion with Bright-rumped Yellow-Finch. Male has an intense yellow body, yellow ear-coverts and large, conical bill. Female has brownish-streaked olive back, but yellow breast and belly, and same comparatively large bill as male. **Habitat** Highland grasslands and high-Andean bogs (bofedales). **Voice** Calls little, but in contact gives a short, repeated monosyllabic *tchip*. We have witnessed a long and elaborate song, reminiscent of other *Sicalis* sp., delivered from a high perch. **Where to see** Visviri and General Lagos (Arica y Parinacota Region).

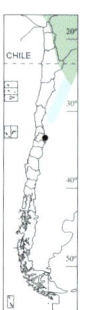

Bright-rumped Yellow-Finch
Sicalis uropygialis R
L 13–14cm. Sexually dimorphic. >4,000m. **ID** Similar to Puna and Greenish Yellow-Finches. Male (even immature) has grey ear-coverts and greyish back. Female has brown back and grey flanks. **Habitat** High-Andean bogs (bofedales), grassland and adjacent rocky areas. **Voice** Not very noisy. Perched or in flight, gives a *twiii - twiii* ... in contact, sometimes repeated. Song is a series of long trills, high but somewhat scratchy, and typical of the genus. **Where to see** Road to Surire near Lauca River, Lejia lagoon (Arica y Parinacota Region), Miscanti and Miñique lagoons (Antofagasta Region).

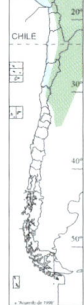

Greenish Yellow-Finch
Sicalis olivascens R
L 13–14cm. Sexually dimorphic. 0–4,000m. **ID** Similar to Greater Yellow-Finch. Male is predominantly greenish-yellow. **Habitat** Diverse environments, but prefers pre-puna areas with scrubby vegetation. **Voice** Calls frequently, giving varied short contact calls ranging from a simple *pweet* to a twittering *trlwit*. Song is also a twitter more than 2 seconds long, repeated many times, at intervals of up to c.12–15 seconds and comprising 25–30 notes, initially ascending, but fading at end (*tllee-tllee-tllee-tllee-tllee-tllee-tllitllee-tllee-tllee-tllee-tlleei-tllee-*...). **Where to see** Zapahuira, Socoroma, Putre (Arica y Parinacota Region), Enquelga/Isluga/Colchane (Tarapacá Region), Machuca and Socaire (Antofagasta Region).

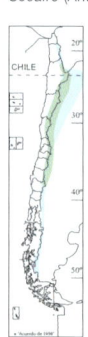

Greater Yellow-Finch
Sicalis auriventris R
L 14–16cm. Sexually dimorphic. 300–4,000m. **ID** Similar to Greenish Yellow-Finch. Male has wholly golden-yellow head. Female is generally pale brown. **Habitat** Mountainous rocky areas and grassland. In winter in thorn and steppe scrub. **Voice** In central Chile, a very common contact call is a somewhat hesitant *twiit*. Song is long, melodious and almost gargling, recalling a small choir (*twee –tegllee-tegllee- tegllee-tegllee*…). **Where to see** El Juncal Andean Park (Valparaíso Region), Farellones and El Yeso Valley (Metropolitan Region).

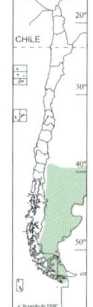

Patagonian Yellow-Finch
Sicalis lebruni R
L 13–14cm. Sexually dimorphic. 0–1,200m. **ID** Similar to other yellow-finches. Male has yellow areas intermixed with brown, and irregular brown patches. Female is mainly brown, with irregular slightly yellow-stained patches. **Habitat** Scattered shrub and rocks. Also Patagonian steppe. **Voice** Not very vocal, but in contact gives clicks (*tchik*), simple whistles (*fwiiit*), composites (*fii-fiiit*) or a slightly drawn-out twitter (*plii-pliw...*). Song given from a high perch mixes whistles, twitters and trills in a long, complex sequence. **Where to see** Pali Ayke National Park, Primera Angostura, Sierra Baguales (Magallanes Region).

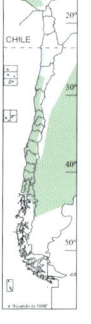

Grassland Yellow-Finch
Sicalis luteola R
L 11–14cm. Sexually dimorphic. 0–1,500m. **ID** Similar to other yellow-finches. Compare female yellow-morph Black-chinned Siskin. Note malar and dark-streaked mantle. Female and juvenile have dark-streaked breast. **Habitat** Grassland and open areas. **Voice** Short and sometimes very quiet *pip* in contact. Song typically in three parts and tends to end as if bird was drowning and singing its last gasp (*twee-twee-twee-twee-twee-twee-trrrrrrrtrrrrrrr-trrrrrrr-tututweet-tututweet-tiii   -tiiiii-tiiiiii-tiiiiiiiiii...*). **Where to see** Rio Maipo wetland (Valparaíso Region), Lampa/Batuco system (Metropolitan Region), coastal plateaux in central Chile (Coquimbo Region–Maule Region).

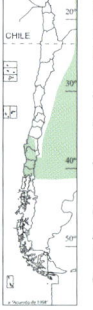

Saffron Finch
Sicalis flaveola R
L 12–14cm. Sexually dimorphic. 0–1,000m. **ID** Male unmistakable. Female similar to other yellow-finches, but is obviously streaked, above and below, and secondaries have yellowish edges. Dark bill and legs. **Habitat** Open areas with shrubby vegetation and scattered trees. **Voice** Most frequently heard in contact is sharp, short and dry *tit*. Song quite elaborate, long and mixes trills, whistles and twitters. A very good singer, and as a result it is a commonly kept cagebird. **Where to see** Victoria, Maullín and other areas in Los Lagos Region.

PLATE 87: SISKINS

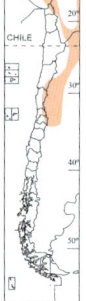

Hooded Siskin
Spinus magellanicus R
L 12–14cm. Sexually dimorphic. 0–4,000m. **ID** Similar to Thick-billed Siskin. Predominantly yellow, with yellow in wing visible when perched, and well-proportioned, sharp-pointed bill. **Habitat** Various environments, from humid to semi-arid, but always in scrub, avoiding dense vegetation. **Voice** Song very similar to congenerics. Usually sings in uncoordinated groups. Contact call differs from other siskins: a monosyllable repeated twice in quick succession, and commonly given many times (*piew-piew ... piew ... piewpiew...*). **Where to see** Lluta Valley, Socoroma, Putre, pre-puna zone in general (Arica y Parinacota Region).

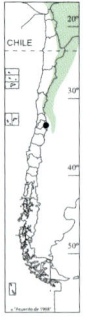

Thick-billed Siskin
Spinus crassirostris R
L 12–14cm. Sexually dimorphic. **N** >4,000m. **C** 2,100–2,600m. **ID** Compare Hooded Siskin and Yellow-rumped Siskin. Black hood reaches throat, below greenish-yellow with whitish-grey flanks, olive-green back and robust conical bill, somewhat disproportionate to bird's overall size. **Habitat** On high plateaus, mainly tied to Queñoa forests (*Polylepis* sp.). In central Chile, shrubby areas in Andes. **Voice** Quieter than congenerics. Song difficult to distinguish from that of sympatric siskins. Most common call in contact is a high trill, which differs from other species of siskin (*tweeee.... tweew.... tweeeee...*). **Where to see** Queñoa forests (Arica y Parinacota Region), Andean Park El Juncal (Valparaíso Region).

Black Siskin
Spinus atratus R
L 12–15cm. Sexes similar. **N** 3,500–4,600m. **ID** Unmistakable. The only almost-black bird in Chilean highlands. Largely black, with yellow flash in wing, on vent and at base of tail. Juvenile is duller black, with smaller yellow patches, more white in wing, and a pale bill. **Habitat** Highlands. Rocky areas with shrubby vegetation, usually near water and high-Andean bogs (bofedales). **Voice** Rarely alone, being usually observed in flocks that call constantly. Song like that of sympatric congenerics. Most commonly heard call in contact is a somewhat hoarse note, interspersed by a long trill, accented at the end (*tcheek ... tcheek ... tweeiiii ...*). **Where to see** Parinacota and Surire Salt Flat (Arica y Parinacota Region), Socaire and Machuca (Antofagasta Region).

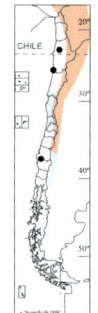

Yellow-rumped Siskin
Spinus uropygialis R
L 12–14cm. Sexually dimorphic. **C** 2,000–4,200m. **ID** Similar to Thick-billed Siskin. Black hood reaches breast in both sexes, with dark back and well-proportioned bill. **Habitat** Mountains with open thickets and grassland. **Voice** Calls constantly. Song like that of sympatric congenerics. In contact regularly gives two trills, separated by a short pause, with the first somewhat stronger, and the second fading (*tuyu-tuiii ... tuyu-tuiii ...*). **Where to see** Farellones and El Yeso Valley (Metropolitan Region).

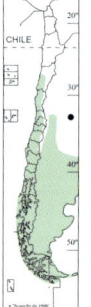

Black-chinned Siskin
Spinus barbatus R
L 12–14cm. Sexually dimorphic. **C** 0–2,600m. **S & A** 0–1,200m. **ID** Female of yellow morph might be confused with female Grassland Yellow-Finch, but note different patterns of yellow on head and yellow in wing of present species, visible at rest and in flight. **Habitat** Shrubby areas and open forests, even city parks and squares. **Voice** Song like that of congenerics. Usually sings in uncoordinated groups. In contact gives a drawn-out twitter, although sometimes drier and shorter (*twiiiiii... tweeet.... twiiiiii...*), and can be repeated over and over. **Where to see** Easy to see throughout its range.

PLATE 88: EGGS OF CHILEAN BIRDS I

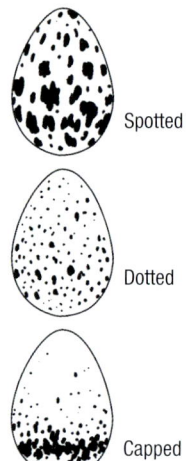

Spotted

Dotted

Capped

1. Rheidae: Lesser Rhea
2. Spheniscidae: Emperor Penguin
3. Diomedeidae: Black-browed Albatross
4. Phoenicopteridae: Chilean Flamingo
5. Pelecanidae: Peruvian Pelican
6. Procellariidae:
   a. Blue Petrel
   b. Peruvian Diving Petrel
7. Hydrobatidae:
   a. White-vented Storm-Petrel
   b. Wilson's Storm-Petrel
8. Phalacrocoracidae: Neotropic Cormorant
9. Sulidae: Peruvian Booby
10. Chionidae: Snowy Sheathbill
11. Phaetontidae: Red-tailed Tropicbird
12. Podicipedidae: Pied-billed Grebe
13. Ardeidae: Black-crowned Night Heron
14. Threskiornithidae: Black-faced Ibis
15. Anatidae:
    a. Black-necked Swan
    b. Ashy-headed Goose
    c. Red Shoveler
    d. Cinnamon Teal
16. Laridae:
    a. Kelp Gull
    b. Grey Gull
17. Sternidae: Peruvian Tern
18. Burhinidae: Peruvian Thick-knee
19. Stercorariidae: Chilean Skua

PLATE 89: EGGS OF CHILEAN BIRDS II

- 20. Rallidae: Red-gartered Coot
- 21. Haematopodidae: American Oystercatcher
- 22. Rostratulidae: South American Painted-snipe
- 23. Charadriidae:
  - a. Diademed Sandpiper-Plover
  - b. Southern Lapwing
  - c. Snowy Plover
- 24. Pluvianellidae: Magellanic Plover
- 25. Recurvirostridae: Black-necked Stilt
- 26. Scolopacidae: Magellanic Snipe
- 27. Thinocoridae: Grey-breasted Seedsnipe
- 28. Cathartidae: Andean Condor
- 29. Falconidae: Striated Caracara
- 30. Accipitridae: Variable Hawk
- 31. Tinamidae:
  - a. Puna Tinamou
  - b. Chilean Tinamou
- 32. Tytonidae: Barn Owl
- 33. Psittacidae: Slender-billed Parakeet
- 34. Rhinocryptidae: Moustached Turca
- 35. Columbidae: Eared Dove
- 36. Caprimulgidae: Band-winged Nightjar
- 37. Odontophoridae: California Quail
- 38. Picidae: Magellanic Woodpecker
- 39. Alcedinidae: Ringed Kingfisher
- 40. Tyrannidae:
  - a. Austral Negrito
  - b. White-crested Elaenia
  - c. Many-coloured Rush-Tyrant
- 41. Hirundinidae: Chilean Swallow
- 42. Furnariidae:
  - a. Wren-like Rushbird
  - b. Common Miner
- 43. Icteridae:
  - a, b, c Shiny Cowbird
  - d. Long-tailed Meadowlark
- 44. Cotingidae: Rufous-tailed Plantcutter
- 45. Turdidae: Austral Thrush
- 46. Mimidae: Chilean Mockingbird
- 47. Trochilidae:
  - a. Green-backed Firecrown
  - b. Chilean Woodstar
- 48. Motacillidae: Correndera Pipit
- 49. Troglodytidae: House Wren
- 50. Thraupidae:
  - a. Grassland Yellow-Finch
  - b. Band-tailed Sierra-Finch
  - c. Common Diuca-Finch
- 51. Emberizidae: Rufous-collared Sparrow
- 52. Fringillidae: Black-chinned Siskin
- 53. Passeridae: House Sparrow

EXTREME VAGRANTS

Atlantic Yellow-nosed Albatross
Thalassarche chlororhynchos   V
L 71–82cm. **W** 180–200cm. Sexes identical. **ID** Could be confused at all ages with Grey-headed Albatross. In adult, head is slightly grey, black sides to bill and largely white underwings with evenly narrow black border. Immature and juvenile have black bill sometimes with slight yellowish on culmen, and same underwing pattern. **Habitat** Temperate and subtropical waters of Atlantic Ocean. **Records** Two records south of Cape Horn, and one in Drake Passage.

Black-winged Petrel
Pterodroma nigripennis   V
L 28–30cm. **W** 63–71cm. Sexes identical. **ID** Unmistakable from below, with heavy black line on otherwise white underwing. From above similar to other *Pterodroma* petrels. **Habitat** Tropical and subtropical waters of central and western Pacific Ocean. **Records** Observed in 2012 and 2013 on Easter Island (Valparaíso Region), and in 2014 on San Félix and San Ambrosio Islands (Valparaíso Region).

Parkinson's Petrel
Procellaria parkinsoni   V
L 41–46cm. **W** 112–123cm. Sexes identical. **ID** Easily confused with White-chinned and Westland Petrels. See comparison of their bills on Plate 9. **Habitat** Temperate, tropical and subtropical waters of south-west to central-east Pacific Ocean. **Records** A few records off northern and far southern Chile. [Alt. Black Petrel]

Red-footed Booby
Sula sula   V
L 66–77cm. **W** 152cm. Sexes identical. **ID** Adult unmistakable by red legs and feet (all morphs). Juvenile/immature can be confused with same ages of Peruvian Booby. **Habitat** Tropical waters. **Records** Observed twice each on Salas y Gómez (1985 and 1997) and (in 2010 and 2013) on Easter Island (Valparaíso Region).

Brown Booby
Sula leucogaster   V
L 64–74cm. **W** 150cm. Sexes similar. **ID** Adult unmistakable. Juvenile and immature can be confused with Peruvian Booby. At all ages has dark brown wing-coverts without whitish edges. **Habitat** Tropical and subtropical waters. **Records** A 2012 record of ssp. *plotus* from Easter Island (Valparaíso Region) and several prior reports (ssp. *brewsteri*) off the Chilean mainland.

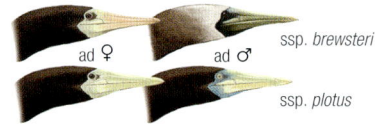

Bristle-thighed Curlew
Numenius tahitiensis   V

L 40–44cm. **W** 90cm. Sexes similar. **ID** Confusion likely with Whimbrel. Overall warmer (ochre or pale cinnamon), especially on rump (with no dark spots). Upperparts marked with large dark spots. **Habitat** Winters on islands in Oceania and breeds in Alaska. **Records** Few records on Easter Island.

Wandering Tattler
Tringa incana   V

L 26–29cm. **W** 66cm. Sexes identical. **ID** Unmistakable. Yellowish legs, somewhat short neck and bill, with grey back, neck and head, and a short white pre-ocular supercilium. **Habitat** Typical of rocky areas on coasts. **Records** Observed in 1999 at Lluta wetland (Arica y Parinacota Region), in 1982 at Antofagasta (Antofagasta Region), in 2013 at Cavancha Peninsula (Antofagasta Region) and several times between 2007 and 2015 on Easter Island (Valparaíso Region).

Buff-breasted Sandpiper
Calidris subruficollis   V

L 18–20cm. **W** 47cm. Sexes identical. **ID** Unmistakable. Mid-sized wader with small, round head, fairly long neck, warm buff face and underparts; and dark brown centres to feathers of upperparts. **Habitat** Grassland areas, not necessarily wet or tied to water. **Records** Observed in 2004 and 2014 at Lluta wetland (Arica y Parinacota Region), and in 2008 near Punta de Choros (Coquimbo Region).

Wilson's Plover
Charadrius wilsonia   V

L 16–20cm. **W** 36cm. Sexes identical. 0–20m. **ID** Unmistakable. Heavy bill that appears disproportionately large for the bird's overall size. **Habitat** Sandy beaches and estuaries. **Records** Observed in 2000 at Lluta wetland (Arica y Parinacota Region) and in 2008 at Loa estuary (Antofagasta Region).

Least Tern
Sternula antillarum   V

L 22–24cm. **W** 51cm. Sexes similar. **ID** Recalls Peruvian Tern. Small size, with shorter bill, different flight pattern, and less white in front of eye. **Habitat** Coastal areas, estuaries and lagoons. **Records** Few records at Lluta wetland, all probably the same returning individual.

Least Bittern
Ixobrychus exilis R?
L 27cm. Sexually dimorphic. 0–500m. **ID** Unmistakable in Chile. Dark back and crown (black in male, brown in female), with streaks confined to breast and belly, and has reddish base to bill in male. **Habitat** Wetlands with reeds (*Scirpus* sp., *Typha* sp.). **Records** Observed in 2004, 2008 and 2016 at Lluta wetland (Arica y Parinacota Region), and in 2016 at Mejillones (Antofagasta Region).

Blue-winged Teal
Spatula discors V
L 38cm. Sexually dimorphic. 0–3,700m. **ID** Male unmistakable. Bluish-grey head with white frontal crescent. Female can be confused with other ducks, but has long dark eyestripe and blue upperwing-coverts (latter also in male). **Habitat** Lakes and lagoons, even those with shallow waters. **Records** Observed in 1965 at Los Vilos (Coquimbo Region), in 1989 at Huechún Reservoir (Metropolitan Region), in 1998 at Lluta wetland (Arica y Parinacota Region) and in 2013 at Pullally (Valparaíso Region).

Dot-winged Crake
Porzana spiloptera V/R?
L 14–15cm. Sexes identical. 0–3,000m. **ID** Could be confused with Black Rail. Small with a brown back and heavy dark streaks (each feather is dark with pale edges) and dense white barring only on flanks and wing-coverts. **Habitat** Freshwater wetlands, with dense vegetation. **Records** A very unusual record at Laguna del Inca (3,000m) in 2012, and a few records since 2014 around Concepción.

Long-winged Harrier
Circus buffoni V
W 119–155cm. Sexually dimorphic. 0–1,000m. **ID** Unmistakable. Long-winged and long-tailed raptor, with heavily barred tail and wings, white rump, dark back, breast-band and wing-coverts, short white supercilium and pale foreface. **Habitat** Wet grassland or near wetlands. **Records** Sporadic records since 1863. Recently, observed in 2000 at Niblinto (Bio-Bio Region), in 2005 at Colina (Metropolitan Region) and in 2006 and 2007 a lone male at Santa Inés (Metropolitan Region).

Swainson's Hawk
Buteo swainsoni V
L 43–55cm. **W** 117–137cm. Sexes similar. 0–2,100m. **ID** Pale morph unmistakable. Intermediate morphs more confusing. Throat and foreneck white, and usually contrast with breast and rest of neck. Dark morph does not always have white throat (which is diagnostic) and could be confused with dark-morph Rufous-tailed Hawk. **Habitat** Mainly in grasslands during non-breeding season in South America (breeds in North America). **Records** In 1869 on Alejandro Selkirk Island, Juan Fernández archipelago (Valparaíso Region, in 1923 at Curicó (Maule Region), in 1994 at Huilma (Los Lagos Region) and in 2007 at Antofagasta (Antofagasta Region).

Brown-chested Martin
Progne tapera   V

L 17–18cm. Sexes identical. 0–1,200m. **ID** Compare the much smaller Bank Swallow (Sand Martin). Broad brown breast-band, some brown spotting on midline of underparts, and pale grey behind ear-coverts. **Habitat** Semi-open areas, with plantations or scattered trees and scrub. **Records** Observed in 1998 at Chacalluta (Arica y Parinacota Region), in 2006 at Matilla (Tarapacá Region) and in 2015 at Baquedano (Antofagasta Region).

Grey-breasted Martin
Progne chalybea   V

L 17–18cm. Sexes identical. 0–1,200m. **ID** Unmistakable. Large, robust-bodied martin; male has mainly steel-blue upperparts, with moderately forked black tail, and mainly whitish-grey underparts, while female is duller, with more mottled upperparts and more typically has grey breast. **Habitat** Wide variety of habitats. Open areas in lowlands, open woodland, grassland and agricultural fields. Also cities. **Records** Toltén estuary in 2010 (Araucanía Region), Copiapo River in 2011 (Atacama Region), Chaca Valley in 2012 (Arica y Parinacota Region) and Humedal Los Batros in 2018 (Bio-Bio Region).

Short-tailed Field-Tyrant
Muscigralla brevicauda   V

L 11–12cm 0–1,500m. Sexes identical. **ID** Unmistakable given combination of very short tail and very long legs. **Habitat** Semi-arid areas with scattered scrub. **Records** Four (in 1970, 1971, 2000 and 2009) all near Arica, in Lluta and Azapa Valleys (Arica y Parinacota Region).

Raimondi's Yellow-Finch
Sicalis raimondii   M/R?

L 11–12cm. Sexually dimorphic. 0–2,500m. **ID** Male somewhat similar to allopatric Bright-rumped Yellow Finch. Pale grey flanks, dirty grey ear-coverts, and greenish-yellow breast and belly. Upperparts pale greyish-brown. Female similar to other yellow-finches. **Habitat** Semi-arid and arid rocky areas with low vegetation, and crop fields. **Records** Considered endemic to Peru. First seen in 2016 at Quebrada de Chiza (Arica y Parinacota Region), where flocks have been reported.

Bobolink
Dolichonyx oryzivorus   V

L 15–18cm. Sexually dimorphic. 0–1,000m. **ID** Unmistakable. A somewhat sparrow-like bird, with a noticeably pale, conical bill, well-streaked upperparts, flanks and lower underparts, and rather stripey head pattern. Superb, mainly black-and-white male in breeding plumage is very unlikely to be seen in Chile. **Habitat** Rice fields, grassland and otherwise humid areas. **Records** Four: in 1968 at Pedro de Valdivia (Antofagasta Region), in 1973 at Molinos (Arica y Parinacota Region) and in 2016 at Ascotán Salt Flat (Antofagasta Region) and Talcahuano (Bio-Bio Region).

APPENDIX I: ENDEMIC BIRD SPECIES IN CHILE

Common Name	Scientific Name
Chilean Tinamou	*Nothoprocta perdicaria*
Juan Fernandez Firecrown	*Sephanoides fernandensis*
Pincoya Storm-Petrel	*Oceanites pincoyae*
Slender-billed Parakeet	*Enicognathus leptorhynchus*
Moustached Turca	*Pteroptochos megapodius*
White-throated Tapaculo	*Scelorchilus albicollis*
Dusky Tapaculo	*Scelorchilus fuscus*
Crag Chilia	*Ochetorhynchus melanurus*
Seaside Cinclodes	*Cinclodes nigrofumosus*
Masafuera Rayadito	*Aphrastura masafuerae*
Dusky-tailed Canastero	*Pseudasthenes humicola*
Juan Fernandez Tit-Tyrant	*Anairetes fernandezianus*

Dusky Tapaculo *Scytalopus fuscus*, Lago Peñuelas National Reserve, Valparaíso Region, Chile (Jose F. Cañas).

APPENDIX II: ENDEMIC BIRD SUBSPECIES IN CHILE

Common Name	Scientific Name
Chilean Tinamou	*Nothoprocta perdicaria perdicaria* *Nothoprocta perdicaria sanborni*
Band-winged Nightjar	*Systellura longirostris mochaensis*
Juan Fernandez Firecrown	*Sephanoides fernandensis fernandensis* *Sephanoides fernandensis leyboldi* (thought to be extinct)
Oasis Hummingbird	*Rhodopis vesper atacamensis*
Plumbeous Rail	*Rallus sanguinolentus landbecki*
Puna Snipe	*Gallinago andina innotata*
Rufous-legged Owl	*Strix rufipes sanborni*
American Kestrel	*Falco sparverius fernandensis*
Burrowing Parrot	*Cyanoliseus patagonus bloxami*
Moustached Turca	*Pteroptochos megapodius megapodius* *Pteroptochos megapodius atacamae*
White-throated Tapaculo	*Scelorchilus albicollis albicollis* *Scelorchilus albicollis atacamae*
Chucao Tapaculo	*Scelorchilus rubecula mochae*
Common Miner	*Geositta cunicularia fissirostris*
Rufous-banded Miner	*Geositta rufipennis harrisoni*
Crag Chilia	*Ochetorhynchus melanurus melanurus* *Ochetorhynchus melanurus atacamae*
Wren-like Rushbird	*Phleocryptes melanops loaensis*
Grey-flanked Cinclodes	*Cinclodes oustaleti baeckstroemii*
Thorn-tailed Rayadito	*Aphrastura spinicauda bullocki* *Aphrastura spinicauda fulva*
Dusky-tailed Canastero	*Pseudasthenes humicola humicola* *Pseudasthenes humicola polysticta* *Pseudasthenes humicola goodalli*
Fire-eyed Diucon	*Xolmis pyrope fortis*
Great Shrike-Tyrant	*Agriornis lividus lividus*
Many-coloured Rush-Tyrant	*Tachuris rubrigastra loaensis*
House Wren	*Troglodytes aedon atacamensis*
Rufous-collared Sparrow	*Zonotrichia capensis antofagastae*
Yellow-winged Blackbird	*Agelasticus thilius thilius*
Grey-hooded Sierra-Finch	*Phrygilus gayi gayi* *Phrygilus gayi minor*
Common Diuca-Finch	*Diuca diuca chiloensis*

Chestnut-throated Huet-huet *Pteroptochos castaneus*, Chile (All Canada Photos/Alamy).

APPENDIX III: THREATENED BIRD SPECIES IN CHILE

All of the threatened Chilean species assessed by the IUCN Red List of Threatened Species are listed here with their threat status: CR (Critically Endangered), EN (Endangered), VU (Vulnerable) and NT (Near Threatened). Sequence follows the South American Classification Committee's (SACC) *List of the Bird Species of South America* (accessed June 2020).

Common Name	Scientific Name	Status
Puna Rhea	*Rhea tarapacensis*	NT
Spectacled Duck	*Speculanas specularis*	NT
Chilean Flamingo	*Phoenicopterus chilensis*	NT
Andean Flamingo	*Phoenicoparrus andinus*	VU
Puna Flamingo	*Phoenicoparrus jamesi*	NT
Northern Silvery Grebe	*Podiceps juninensis*	NT
Hooded Grebe	*Podiceps gallardoi*	CR
Chimney Swift	*Chaetura pelagica*	VU
Juan Fernandez Firecrown	*Sephanoides fernandensis*	CR
Chilean Woodstar	*Eulidia yarrellii*	CR
Austral Rail	*Rallus antarcticus*	VU
Black Rail	*Laterallus jamaicensis*	EN
Dot-winged Crake	*Porzana spiloptera*	VU
Horned Coot	*Fulica cornuta*	NT
Snowy Plover	*Charadrius nivosus*	NT
Diademed Sandpiper-Plover	*Phegornis mitchellii*	NT
Magellanic Plover	*Pluvianellus socialis*	NT
Bristle-thighed Curlew	*Numenius tahitiensis*	VU
Red Knot	*Calidris canutus*	NT
Buff-breasted Sandpiper	*Calidris subruficollis*	NT
Semipalmated Sandpiper	*Calidris pusilla*	NT
Fuegian Snipe	*Gallinago stricklandii*	NT
Peruvian Tern	*Sternula lorata*	EN
Inca Tern	*Larosterna inca*	NT
Elegant Tern	*Thalasseus elegans*	NT
Emperor Penguin	*Aptenodytes forsteri*	NT
Humboldt Penguin	*Spheniscus humboldti*	VU
Magellanic Penguin	*Spheniscus magellanicus*	NT
Macaroni Penguin	*Eudyptes chrysolophus*	VU
Southern Rockhopper Penguin	*Eudyptes chrysocome*	VU

Common Name	Scientific Name	Status
Waved Albatross	*Phoebastria irrorata*	CR
Northern Royal Albatross	*Diomedea sanfordi*	EN
Southern Royal Albatross	*Diomedea epomophora*	VU
Wandering Albatross	*Diomedea exulans*	VU
Antipodean Albatross	*Diomedea antipodensis*	EN
Light-mantled Albatross	*Phoebetria palpebrata*	NT
Atlantic Yellow-nosed Albatross	*Thalassarche chlororhynchos*	EN
Grey-headed Albatross	*Thalassarche chrysostoma*	EN
Buller's Albatross	*Thalassarche bulleri bulleri*	NT
Pacific Albatross	*Thalassarche bulleri platei*	NT
White-capped Albatross	*Thalassarche steadi*	NT
Salvin's Albatross	*Thalassarche salvini*	VU
Chatham Albatross	*Thalassarche eremita*	VU
Polynesian Storm-Petrel	*Nesofregetta fuliginosa*	EN
White-faced Storm-Petrel	*Pelagodroma marina*	EN
Markham's Storm-Petrel	*Oceanodroma markhami*	NT
Ringed Storm-Petrel	*Oceanodroma hornbyi*	NT
Atlantic Petrel	*Pterodroma incerta*	EN
Masatierra Petrel	*Pterodroma defilippiana*	VU
Stejneger's Petrel	*Pterodroma longirostris*	VU
Henderson Petrel	*Pterodroma atrata*	EN
Phoenix Petrel	*Pterodroma alba*	EN
Mottled Petrel	*Pterodroma inexpectata*	NT
Juan Fernandez Petrel	*Pterodroma externa*	VU
Grey Petrel	*Procellaria cinerea*	NT
White-chinned Petrel	*Procellaria aequinoctialis*	VU
Parkinson's Petrel	*Procellaria parkinsoni*	VU
Westland Petrel	*Procellaria westlandica*	EN
Buller's Shearwater	*Ardenna bulleri*	VU
Sooty Shearwater	*Ardenna grisea*	NT
Pink-footed Shearwater	*Ardenna creatopus*	VU
Peruvian Diving Petrel	*Pelecanoides garnoti*	EN
Red-legged Cormorant	*Phalacrocorax gaimardi*	NT

Common Name	Scientific Name	Status
Guanay Cormorant	*Phalacrocorax bougainvilliiorum*	NT
Peruvian Pelican	*Pelecanus thagus*	NT
Andean Ibis	*Theristicus branickii*	NT
Andean Condor	*Vultur gryphus*	NT
Rufous-tailed Hawk	*Buteo ventralis*	VU
Striated Caracara	*Phalcoboenus australis*	NT
Red-masked Parakeet	*Psittacara erythrogenys*	NT
Black Cinclodes	*Cinclodes maculirostris*	NT
Masafuera Rayadito	*Aphrastura masafuerae*	CR
Juan Fernandez Tit-Tyrant	*Anairetes fernandezianus*	NT
White-tailed Shrike-Tyrant	*Agriornis albicauda*	VU
Peruvian Martin	*Progne murphyi*	VU
Tamarugo Conebill	*Conirostrum tamarugense*	VU
Slender-billed Finch	*Xenospingus concolor*	NT

White-throated Tapaculo *Scelorchilus albicollis*, Llanos de Challe National Park, Atacama Region, Chile (Jose F. Cañas).

CHECKLIST OF THE BIRDS OF CHILE

This checklist includes all the species covered in this book, using the same taxonomy as in the species accounts. Sequence follows the South American Classification Committee's (SACC) List of the Bird Species of South America (accessed June 2020). Key to the status codes: R = Resident; M = Migrant visitor; V = Vagrant; e = endemic; i = introduced; ? = uncertain status (see pages 13–14 for a fuller explanation).

	RHEIDAE (RHEAS)		
☐	Puna Rhea	*Rhea tarapacensis*	R
☐	Lesser Rhea	*Rhea pennata*	R
	TINAMIDAE (TINAMOUS)		
☐	Ornate Tinamou	*Nothoprocta ornata*	R
☐	Chilean Tinamou	*Nothoprocta perdicaria*	Re
☐	Elegant Crested Tinamou	*Eudromia elegans*	R/V?
☐	Puna Tinamou	*Tinamotis pentlandii*	R
☐	Patagonian Tinamou	*Tinamotis ingoufi*	R/V?
	ANATIDAE (DUCKS, GEESE AND SWANS)		
☐	Fulvous Whistling Duck	*Dendrocygna bicolor*	V
☐	White-faced Whistling Duck	*Dendrocygna viduata*	V
☐	Black-bellied Whistling Duck	*Dendrocygna autumnalis*	V
☐	Black-necked Swan	*Cygnus melancoryphus*	R
☐	Coscoroba Swan	*Coscoroba coscoroba*	R
☐	Andean Goose	*Chloephaga melanopterus*	R
☐	Upland Goose	*Chloephaga picta*	R
☐	Kelp Goose	*Chloephaga hybrida*	R
☐	Ashy-headed Goose	*Chloephaga poliocephala*	R
☐	Ruddy-headed Goose	*Chloephaga rubidiceps*	R
☐	Torrent Duck	*Merganetta armata*	R
☐	Flying Steamerduck	*Tachyeres patachonicus*	R
☐	Magellanic Steamerduck	*Tachyeres pteneres*	R
☐	Crested Duck	*Lophonetta specularioides*	R
☐	Spectacled Duck	*Speculanas specularis*	R
☐	Puna Teal	*Spatula puna*	R
☐	Silver Teal	*Spatula versicolor*	R
☐	Red Shoveler	*Spatula platalea*	R
☐	Blue-winged Teal	*Spatula discors*	V
☐	Cinnamon Teal	*Spatula cyanoptera*	R
☐	Chiloe Wigeon	*Mareca sibilatrix*	R
☐	White-cheeked Pintail	*Anas bahamensis*	R
☐	Yellow-billed Pintail	*Anas georgica*	R

☐	Yellow-billed Teal	*Anas flavirostris*	R
☐	Rosy-billed Pochard	*Netta peposaca*	R
☐	Black-headed Duck	*Heteronetta atricapilla*	R
☐	Andean Duck	*Oxyura ferruginea*	R
☐	Lake Duck	*Oxyura vitatta*	R
	ODONTOPHORIDAE (NEW WORLD QUAILS)		
☐	California Quail	*Callipepla californica*	Ri
	PHASIANIDAE (PHEASANTS)		
☐	Common Pheasant	*Phasianus colchicus*	Ri
	PHOENICOPTERIDAE (FLAMINGOS)		
☐	Chilean Flamingo	*Phoenicopterus chilensis*	R
☐	Andean Flamingo	*Phoenicoparrus andinus*	R
☐	Puna Flamingo	*Phoenicoparrus jamesi*	R
	PODICIPEDIDAE (GREBES)		
☐	White-tufted Grebe	*Rollandia rolland*	R
☐	Pied-billed Grebe	*Podilymbus podiceps*	R
☐	Great Grebe	*Podiceps major*	R
☐	Northern Silvery Grebe	*Podiceps juninensis*	R
☐	Southern Silvery Grebe	*Podiceps occipitalis*	R
☐	Hooded Grebe	*Podiceps gallardoi*	V
	COLUMBIDAE (PIGEONS AND DOVES)		
☐	Rock Dove/Feral Pigeon	*Columba livia*	Ri
☐	White-winged Pigeon	*Patagioenas albipennis*	M/R?
☐	Chilean Pigeon	*Patagioenas araucana*	R
☐	West Peruvian Dove	*Zenaida meloda*	R
☐	Eared Dove	*Zenaida auriculata*	R
☐	Bare-faced Ground-Dove	*Metriopelia ceciliae*	R
☐	Black-winged Ground-Dove	*Metriopelia melanoptera*	R
☐	Golden-spotted Ground-Dove	*Metriopelia aymara*	R
☐	Ruddy Ground-Dove	*Columbina talpacoti*	V
☐	Picui Ground-Dove	*Columbina picui*	R
☐	Croaking Ground-Dove	*Columbina cruziana*	R
	CUCULIDAE (CUCKOOS)		
☐	Groove-billed Ani	*Crotophaga sulcirostris*	R
	CAPRIMULGIDAE (NIGHTJARS)		
☐	Lesser Nighthawk	*Chordeiles acutipennis*	V
☐	Band-winged Nightjar	*Systellura longirostris*	R
☐	Tschudi's Nightjar	*Systellura decussata*	R

	APODIDAE (SWIFTS)		
☐	Chimney Swift	*Chaetura pelagica*	V
☐	Andean Swift	*Aeronautes andecolus*	M/R?
	TROCHILIDAE (HUMMINGBIRDS)		
☐	Sparkling Violetear	*Colibri coruscans*	R
☐	Green-backed Firecrown	*Sephanoides sephaniodes*	R
☐	Juan Fernandez Firecrown	*Sephanoides fernandensis*	Re
☐	Andean Hillstar	*Oreotrochilus estella*	R
☐	White-sided Hillstar	*Oreotrochilus leucopleurus*	R
☐	Giant Hummingbird	*Patagona gigas*	R
☐	Chilean Woodstar	*Eulidia yarrellii*	R
☐	Oasis Hummingbird	*Rhodopis vesper*	R
☐	Peruvian Sheartail	*Thaumastura cora*	R
	RALLIDAE (RAILS, CRAKES AND GALLINULES)		
☐	Austral Rail	*Rallus antarcticus*	R
☐	Purple Gallinule	*Porphyrio martinicus*	V
☐	Black Rail	*Laterallus jamaicensis*	R
☐	Plumbeous Rail	*Pardirallus sanguinolentus*	R
☐	Dot-winged Crake	*Porzana spiloptera*	V/R?
☐	Spot-flanked Gallinule	*Gallinula melanops*	R
☐	Common Gallinule	*Gallinula galeata*	R
☐	Red-fronted Coot	*Fulica rufifrons*	R
☐	Horned Coot	*Fulica cornuta*	R
☐	Giant Coot	*Fulica gigantea*	R
☐	Red-gartered Coot	*Fulica armillata*	R
☐	Andean Coot	*Fulica ardesiaca*	R
☐	White-winged Coot	*Fulica leucoptera*	R
	CHARADRIIDAE (PLOVERS AND LAPWINGS)		
☐	American Golden Plover	*Pluvialis dominica*	M
☐	Grey Plover	*Pluvialis squatarola*	M
☐	Tawny-throated Dotterel	*Oreopholus ruficollis*	R
☐	Southern Lapwing	*Vanellus chilensis*	R
☐	Andean Lapwing	*Vanellus resplendens*	R
☐	Rufous-chested Dotterel	*Charadrius modestus*	R
☐	Killdeer	*Charadrius vociferus*	R
☐	Wilson's Plover	*Charadrius wilsonia*	V
☐	Collared Plover	*Charadrius collaris*	R
☐	Puna Plover	*Charadrius alticola*	R

☐	Two-banded Plover	*Charadrius falklandicus*	R
☐	Snowy Plover	*Charadrius nivosus*	R
☐	Diademed Sandpiper-Plover	*Phegornis mitchellii*	R
	HAEMATOPODIDAE (OYSTERCATCHERS)		
☐	American Oystercatcher	*Haematopus palliatus*	R
☐	Blackish Oystercatcher	*Haematopus ater*	R
☐	Magellanic Oystercatcher	*Haematopus leucopodus*	R
	RECURVIROSTRIDAE (STILTS AND AVOCETS)		
☐	Black-necked Stilt	*Himantopus mexicanus*	R
☐	Andean Avocet	*Recurvirostra andina*	R
	BURHINIDAE (THICK-KNEES)		
☐	Peruvian Thick-knee	*Burhinus superciliaris*	R
	CHIONIDAE (SHEATHBILLS)		
☐	Snowy Sheathbill	*Chionis albus*	R
	PLUVIANELLIDAE (MAGELLANIC PLOVER)		
☐	Magellanic Plover	*Pluvianellus socialis*	R
	SCOLOPACIDAE (SANDPIPERS AND ALLIES)		
☐	Upland Sandpiper	*Bartramia longicauda*	V/M?
☐	Bristle-thighed Curlew	*Numenius tahitiensis*	V
☐	Whimbrel	*Numenius phaeopus*	M
☐	Hudsonian Godwit	*Limosa haemastica*	M
☐	Marbled Godwit	*Limosa fedoa*	V
☐	Ruddy Turnstone	*Arenaria interpres*	M
☐	Red Knot	*Calidris canutus*	M
☐	Surfbird	*Calidris virgata*	M
☐	Stilt Sandpiper	*Calidris himantopus*	M
☐	Sanderling	*Calidris alba*	M
☐	Baird's Sandpiper	*Calidris bairdii*	M
☐	Least Sandpiper	*Calidris minutilla*	V
☐	White-rumped Sandpiper	*Calidris fuscicollis*	M
☐	Buff-breasted Sandpiper	*Calidris subruficollis*	V
☐	Pectoral Sandpiper	*Calidris melanotos*	M
☐	Semipalmated Sandpiper	*Calidris pusilla*	M
☐	Western Sandpiper	*Calidris mauri*	V
☐	Short-billed Dowitcher	*Limnodromus griseus*	V
☐	Fuegian Snipe	*Gallinago stricklandii*	R
☐	Magellanic Snipe	*Gallinago magellanica*	R
☐	Puna Snipe	*Gallinago andina*	R

☐	Wilson's Phalarope	*Steganopus tricolor*	M
☐	Red-necked Phalarope	*Phalaropus lobatus*	M
☐	Red Phalarope	*Phalaropus fulicarius*	M
☐	Spotted Sandpiper	*Actitis macularius*	M
☐	Solitary Sandpiper	*Tringa solitaria*	V
☐	Wandering Tattler	*Tringa incana*	V
☐	Greater Yellowlegs	*Tringa melanoleuca*	M
☐	Willet	*Tringa semipalmata*	M
☐	Lesser Yellowlegs	*Tringa flavipes*	M
	THINOCORIDAE (SEEDSNIPES)		
☐	Rufous-bellied Seedsnipe	*Attagis gayi*	R
☐	White-bellied Seedsnipe	*Attagis malouinus*	R
☐	Grey-breasted Seedsnipe	*Thinocorus orbignyianus*	R
☐	Least Seedsnipe	*Thinocorus rumicivorus*	R
	ROSTRATULIDAE (PAINTED-SNIPES)		
☐	South American Painted-snipe	*Nycticryphes semicollaris*	R
	STERCORARIIDAE (JAEGERS AND SKUAS)		
☐	Chilean Skua	*Stercorarius chilensis*	R
☐	South Polar Skua	*Stercorarius maccormicki*	R
☐	Brown Skua	*Stercorarius antarcticus*	R
☐	Pomarine Jaeger	*Stercorarius pomarinus*	M
☐	Parasitic Jaeger	*Stercorarius parasiticus*	M
☐	Long-tailed Jaeger	*Stercorarius longicaudus*	M
	RYNCHOPIDAE (SKIMMERS)		
☐	Black Skimmer	*Rynchops niger*	M
	LARIDAE (GULLS AND TERNS)		
☐	Swallow-tailed Gull	*Creagrus furcatus*	V
☐	Sabine's Gull	*Xema sabini*	V
☐	Andean Gull	*Chroicocephalus serranus*	R
☐	Brown-hooded Gull	*Chroicocephalus maculipennis*	R
☐	Grey-hooded Gull	*Chroicocephalus cirrocephalus*	V
☐	Dolphin Gull	*Leucophaeus scoresbii*	R
☐	Grey Gull	*Leucophaeus modestus*	R
☐	Laughing Gull	*Leucophaeus atricilla*	V
☐	Franklin's Gull	*Leucophaeus pipixcan*	M
☐	Belcher's Gull	*Larus belcheri*	R
☐	Kelp Gull	*Larus dominicanus*	R
☐	Brown Noddy	*Anous stolidus*	R

☐	Grey Noddy	*Anous albivitta*	R
☐	White Tern	*Gygis alba*	R/M?
☐	Sooty Tern	*Onychoprion fuscatus*	R
☐	Grey-backed Tern	*Onychoprion lunatus*	R
☐	Least Tern	*Sternula antillarum*	V
☐	Peruvian Tern	*Sternula lorata*	R
☐	Inca Tern	*Larosterna inca*	R
☐	Common Tern	*Sterna hirundo*	M
☐	Arctic Tern	*Sterna paradisaea*	M
☐	South American Tern	*Sterna hirundinacea*	R
☐	Antarctic Tern	*Sterna vittata*	R
☐	Snowy-crowned Tern	*Sterna trudeaui*	R
☐	Elegant Tern	*Thalasseus elegans*	M
☐	Sandwich Tern	*Thalasseus sandvicensis*	V
	PHAETHONTIDAE (TROPICBIRDS)		
☐	Red-billed Tropicbird	*Phaethon aethereus*	R
☐	Red-tailed Tropicbird	*Phaethon rubricauda*	R
☐	White-tailed Tropicbird	*Phaethon lepturus*	R
	SPHENISCIDAE (PENGUINS)		
☐	King Penguin	*Aptenodytes patagonicus*	R
☐	Emperor Penguin	*Aptenodytes forsteri*	R
☐	Gentoo Penguin	*Pygoscelis papua*	R
☐	Adelie Penguin	*Pygoscelis adeliae*	R
☐	Chinstrap Penguin	*Pygoscelis antarcticus*	R
☐	Little Penguin	*Eudyptula minor*	V
☐	Humboldt Penguin	*Spheniscus humboldti*	R
☐	Magellanic Penguin	*Spheniscus magellanicus*	R
☐	Macaroni Penguin	*Eudyptes chrysolophus*	R
☐	Southern Rockhopper Penguin	*Eudyptes chrysocome*	R
	DIOMEDEIDAE (ALBATROSSES)		
☐	Waved Albatross	*Phoebastria irrorata*	V
☐	Northern Royal Albatross	*Diomedea sanfordi*	M
☐	Southern Royal Albatross	*Diomedea epomophora*	M
☐	Wandering Albatross	*Diomedea exulans*	M
☐	Antipodean Albatross	*Diomedea antipodensis*	M
☐	Light-mantled Albatross	*Phoebetria palpebrata*	M
☐	Atlantic Yellow-nosed Albatross	*Thalassarche chlororhynchos*	V
☐	Black-browed Albatross	*Thalassarche melanophris*	R

☐	Grey-headed Albatross	*Thalassarche chrysostoma*	R
☐	Buller's Albatross	*Thalassarche bulleri bulleri*	M
☐	Pacific Albatross	*Thalassarche bulleri platei*	M
☐	White-capped Albatross	*Thalassarche steadi*	V
☐	Salvin's Albatross	*Thalassarche salvini*	M
☐	Chatham Albatross	*Thalassarche eremita*	M
	OCEANITIDAE (SOUTHERN STORM-PETRELS)		
☐	White-bellied Storm-Petrel	*Fregetta grallaria*	R
☐	Black-bellied Storm-Petrel	*Fregetta tropica*	M
☐	Polynesian Storm-Petrel	*Nesofregetta fuliginosa*	R
☐	Wilson's Storm-Petrel	*Oceanites oceanicus*	R
☐	Pincoya Storm-Petrel	*Oceanites pincoyae*	Re?
☐	White-vented Storm-Petrel	*Oceanites gracilis*	R
☐	Grey-backed Storm-Petrel	*Garrodia nereis*	M
☐	White-faced Storm-Petrel	*Pelagodroma marina*	M
	HYDROBATIDAE (NORTHERN STORM-PETRELS)		
☐	Wedge-rumped Storm-Petrel	*Oceanodroma tethys*	R
☐	Markham's Storm-Petrel	*Oceanodroma markhami*	R
☐	Ringed Storm-Petrel	*Oceanodroma hornbyi*	R
	PROCELLARIIDAE (PETRELS AND SHEARWATERS)		
☐	Southern Giant Petrel	*Macronectes giganteus*	R
☐	Northern Giant Petrel	*Macronectes halli*	M
☐	Southern Fulmar	*Fulmarus glacialoides*	R
☐	Antarctic Petrel	*Thalassoica antarctica*	R
☐	Cape Petrel	*Daption capense*	R
☐	Snow Petrel	*Pagodroma nivea*	R
☐	Kerguelen Petrel	*Aphrodroma brevirostris*	V
☐	Atlantic Petrel	*Pterodroma incerta*	V
☐	White-headed Petrel	*Pterodroma lessonii*	M
☐	Cook's Petrel	*Pterodroma cookii*	M
☐	Black-winged Petrel	*Pterodroma nigripennis*	V
☐	Masatierra Petrel	*Pterodroma defilippiana*	R
☐	Stejneger's Petrel	*Pterodroma longirostris*	R
☐	Murphy's Petrel	*Pterodroma ultima*	R
☐	Kermadec Petrel	*Pterodroma neglecta*	R
☐	Herald Petrel	*Pterodroma heraldica*	R
☐	Henderson Petrel	*Pterodroma atrata*	M/R?
☐	Phoenix Petrel	*Pterodroma alba*	R

☐	Mottled Petrel	*Pterodroma inexpectata*	V/M?
☐	Juan Fernandez Petrel	*Pterodroma externa*	R
☐	Blue Petrel	*Halobaena caerulea*	R
☐	Antarctic Prion	*Pachyptila desolata*	M
☐	Slender-billed Prion	*Pachyptila belcheri*	R
☐	Grey Petrel	*Procellaria cinerca*	M
☐	White-chinned Petrel	*Procellaria aequinoctialis*	M
☐	Parkinson's Petrel	*Procellaria parkinsoni*	V
☐	Westland Petrel	*Procellaria westlandica*	M
☐	Wedge-tailed Shearwater	*Ardenna pacifica*	R
☐	Buller's Shearwater	*Ardenna bulleri*	M
☐	Sooty Shearwater	*Ardenna grisea*	R
☐	Great Shearwater	*Ardenna gravis*	M/V
☐	Pink-footed Shearwater	*Ardenna creatopus*	R
☐	Christmas Shearwater	*Puffinus nativitatis*	R
☐	Manx Shearwater	*Puffinus puffinus*	V
☐	Subantarctic Shearwater	*Puffinus elegans*	V
☐	Peruvian Diving Petrel	*Pelecanoides garnoti*	R
☐	Common Diving Petrel	*Pelecanoides urinatrix*	R
☐	Magellanic Diving Petrel	*Pelecanoides magellani*	R
	FREGATIDAE (FRIGATEBIRDS)		
☐	Great Frigatebird	*Fregata minor*	R
	SULIDAE (BOOBIES)		
☐	Blue-footed Booby	*Sula nebouxii*	R
☐	Peruvian Booby	*Sula variegata*	R
☐	Masked Booby	*Sula dactylatra*	R
☐	Red-footed Booby	*Sula sula*	V
☐	Brown Booby	*Sula leucogaster*	V
	PHALACROCORACIDAE (CORMORANTS)		
☐	Red-legged Cormorant	*Phalacrocorax gaimardi*	R
☐	Neotropic Cormorant	*Phalacrocorax brasilianus*	R
☐	Magellanic Cormorant	*Phalacrocorax magellanicus*	R
☐	Guanay Cormorant	*Phalacrocorax bougainvilliiorum*	R
☐	Imperial Cormorant	*Phalacrocorax atriceps*	R
☐	Antarctic Cormorant	*Phalacrocorax bransfieldensis*	R
	PELECANIDAE (PELICANS)		
☐	Brown Pelican	*Pelecanus occidentalis*	V
☐	Peruvian Pelican	*Pelecanus thagus*	R

	ARDEIDAE (HERONS AND BITTERNS)		
☐	Least Bittern	*Ixobrychus exilis*	R?
☐	Striped-backed Bittern	*Ixobrychus involucris*	R
☐	Black-crowned Night Heron	*Nycticorax nycticorax*	R
☐	Yellow-crowned Night Heron	*Nyctanassa violacea*	V
☐	Green-backed Heron	*Butorides striata*	V
☐	Cattle Egret	*Bubulcus ibis*	R
☐	Cocoi Heron	*Ardea cocoi*	R
☐	Great Egret	*Ardea alba*	R
☐	Tricoloured Heron	*Egretta tricolor*	M
☐	Snowy Egret	*Egretta thula*	R
☐	Little Blue Heron	*Egretta caerulea*	M/R?
	THRESKIORNITHIDAE (IBISES AND SPOONBILLS)		
☐	Puna Ibis	*Plegadis ridgwayi*	M
☐	Andean Ibis	*Theristicus branickii*	M/V?
☐	Black-faced Ibis	*Theristicus melanopis*	R
☐	Roseate Spoonbill	*Platalea ajaja*	V
	CATHARTIDAE (NEW WORLD VULTURES)		
☐	Andean Condor	*Vultur gryphus*	R
☐	Black Vulture	*Coragyps atratus*	R
☐	Turkey Vulture	*Cathartes aura*	R
	PANDIONIDAE (OSPREY)		
☐	Osprey	*Pandion haliaetus*	M
	ACCIPITRIDAE (HAWKS, HARRIERS AND KITES)		
☐	White-tailed Kite	*Elanus leucurus*	R
☐	Cinereous Harrier	*Circus cinereus*	R
☐	Long-winged Harrier	*Circus buffoni*	V
☐	Chilean Hawk	*Accipiter chilensis*	R
☐	Harris's Hawk	*Parabuteo unicinctus*	R
☐	Variable Hawk	*Geranoaetus polyosoma*	R
☐	Puna Hawk	*Geranoaetus polyosoma poecilochrous*	R
☐	Masafuera Hawk	*Geranoaetus polyosoma exsul*	R
☐	Black-chested Buzzard-Eagle	*Geranoaetus melanoleucus*	R
☐	White-throated Hawk	*Buteo albigula*	R
☐	Swainson's Hawk	*Buteo swainsoni*	V
☐	Rufous-tailed Hawk	*Buteo ventralis*	R
	TYTONIDAE (BARN OWLS)		
☐	Barn Owl	*Tyto alba*	R

	STRIGIDAE (OWLS)		
☐	Magellanic Horned Owl	*Bubo magellanicus*	R
☐	Rufous-legged Owl	*Strix rufipes*	R
☐	Peruvian Pygmy-Owl	*Glaucidium peruanum*	R
☐	Austral Pygmy-Owl	*Glaucidium nana*	R
☐	Burrowing Owl	*Athene cunicularia*	R
☐	Short-eared Owl	*Asio flammeus*	R
	ALCEDINIDAE (KINGFISHERS)		
☐	Ringed Kingfisher	*Megaceryle torquata*	R
	PICIDAE (WOODPECKERS)		
☐	Striped Woodpecker	*Veniliornis lignarius*	R
☐	Magellanic Woodpecker	*Campephilus magellanicus*	R
☐	Chilean Flicker	*Colaptes pitius*	R
☐	Andean Flicker	*Colaptes rupicola*	R
	FALCONIDAE (FALCONS AND CARACARAS)		
☐	Southern Caracara	*Caracara plancus*	R
☐	Mountain Caracara	*Phalcoboenus megalopterus*	R
☐	White-throated Caracara	*Phalcoboenus albogularis*	R
☐	Striated Caracara	*Phalcoboenus australis*	R
☐	Chimango Caracara	*Milvago chimango*	R
☐	American Kestrel	*Falco sparverius*	R
☐	Aplomado Falcon	*Falco femoralis*	R
☐	Peregrine Falcon	*Falco peregrinus*	R
	PSITTACIDAE (NEW WORLD PARROTS)		
☐	Mountain Parakeet	*Psilopsiagon aurifrons*	R
☐	Monk Parakeet	*Myiopsitta monachus*	Ri
☐	Austral Parakeet	*Enicognathus ferrugineus*	R
☐	Slender-billed Parakeet	*Enicognathus leptorhynchus*	Re
☐	Burrowing Parrot	*Cyanoliseus patagonus*	R
☐	Red-masked Parakeet	*Psittacara erythrogenys*	Ri?
	RHINOCRYPTIDAE (TAPACULOS)		
☐	Chestnut-throated Huet-huet	*Pteroptochos castaneus*	R
☐	Black-throated Huet-huet	*Pteroptochos tarnii*	R
☐	Moustached Turca	*Pteroptochos megapodius*	Re
☐	White-throated Tapaculo	*Scelorchilus albicollis*	Re
☐	Chucao Tapaculo	*Scelorchilus rubecula*	R
☐	Ochre-flanked Tapaculo	*Eugralla paradoxa*	R
☐	Magellanic Tapaculo	*Scytalopus magellanicus*	R
☐	Dusky Tapaculo	*Scytalopus fuscus*	Re

	FURNARIIDAE (OVENBIRDS)		
☐	Common Miner	*Geositta cunicularia*	R
☐	Puna Miner	*Geositta punensis*	R
☐	Rufous-banded Miner	*Geositta rufipennis*	R
☐	Greyish Miner	*Geositta maritima*	R
☐	Short-billed Miner	*Geositta antarctica*	R
☐	Creamy-rumped Miner	*Geositta isabellina*	R
☐	White-throated Treerunner	*Pygarrhichas albogularis*	R
☐	Straight-billed Earthcreeper	*Ochetorhynchus ruficaudus*	R
☐	Band-tailed Earthcreeper	*Ochetorhynchus phoenicurus*	R
☐	Crag Chilia	*Ochetorhynchus melanurus*	Re
☐	Wren-like Rushbird	*Phleocryptes melanops*	R
☐	Patagonian Forest Earthcreeper	*Upucerthia saturatior*	R
☐	Scale-throated Earthcreeper	*Upucerthia dumetaria*	R
☐	White-throated Earthcreeper	*Upucerthia albigula*	R
☐	Buff-breasted Earthcreeper	*Upucerthia validirostris*	R
☐	Buff-winged Cinclodes	*Cinclodes fuscus*	R
☐	Black Cinclodes	*Cinclodes maculirostris*	R
☐	Cream-winged Cinclodes	*Cinclodes albiventris*	R
☐	Grey-flanked Cinclodes	*Cinclodes oustaleti*	R
☐	White-winged Cinclodes	*Cinclodes atacamensis*	R
☐	Dark-bellied Cinclodes	*Cinclodes patagonicus*	R
☐	Seaside Cinclodes	*Cinclodes nigrofumosus*	Re
☐	Thorn-tailed Rayadito	*Aphrastura spinicauda*	R
☐	Masafuera Rayadito	*Aphrastura masafuerae*	Re
☐	Des Murs's Wiretail	*Sylviorthorhynchus desmursii*	R
☐	Plain-mantled Tit-Spinetail	*Leptasthenura aegithaloides*	R
☐	Puna Tit-Spinetail	*Leptasthenura berlepschi*	R
☐	Patagonian Tit-Spinetail	*Leptasthenura pallida*	R
☐	Streaked Tit-Spinetail	*Leptasthenura striata*	R
☐	Arequipa Canastero	*Asthenes arequipae*	R
☐	Austral Canastero	*Asthenes anthoides*	R
☐	Cordilleran Canastero	*Asthenes modesta*	R
☐	Sharp-billed Canastero	*Asthenes pyrrholeuca*	R
☐	Canyon Canastero	*Asthenes pudibunda*	R
☐	Dusky-tailed Canastero	*Pseudasthenes humicola*	Re
	COTINGIDAE (COTINGAS)		
☐	Rufous-tailed Plantcutter	*Phytotoma rara*	R

	TYRANNIDAE (TYRANT FLYCATCHERS)		
☐	White-crested Elaenia	*Elaenia albiceps*	R
☐	Pied-crested Tit-Tyrant	*Anairetes reguloides*	R
☐	Yellow-billed Tit-Tyrant	*Anairetes flavirostris*	R
☐	Tufted Tit-Tyrant	*Anairetes parulus*	R
☐	Juan Fernandez Tit-Tyrant	*Anairetes fernandezianus*	Re
☐	Ticking Doradito	*Pseudocolopteryx citreola*	R
☐	Short-tailed Field-Tyrant	*Muscigralla brevicauda*	V
☐	Great Kiskadee	*Pitangus sulphuratus*	V
☐	Tropical Kingbird	*Tyrannus melancholicus*	V
☐	Fork-tailed Flycatcher	*Tyrannus savana*	V
☐	Eastern Kingbird	*Tyrannus tyrannus*	V
☐	Rufescent Flycatcher	*Myiophobus rufescens*	R
☐	Patagonian Tyrant	*Colorhamphus parvirostris*	R
☐	D'Orbigny's Chat-Tyrant	*Ochthoeca oenanthoides*	R
☐	White-browed Chat-Tyrant	*Ochthoeca leucophrys*	R
☐	Vermilion Flycatcher	*Pyrocephalus rubinus*	R
☐	Austral Negrito	*Lessonia rufa*	R
☐	Andean Negrito	*Lessonia oreas*	R
☐	Spectacled Tyrant	*Hymenops perspicillatus*	R
☐	Spot-billed Ground-Tyrant	*Muscisaxicola maculirostris*	R
☐	Puna Ground-Tyrant	*Muscisaxicola juninensis*	R
☐	Cinereous Ground-Tyrant	*Muscisaxicola cinereus*	R
☐	White-fronted Ground-Tyrant	*Muscisaxicola albifrons*	R
☐	Ochre-naped Ground-Tyrant	*Muscisaxicola flavinucha*	R
☐	Rufous-naped Ground-Tyrant	*Muscisaxicola rufivertex*	R
☐	Dark-faced Ground-Tyrant	*Muscisaxicola maclovianus*	R
☐	White-browed Ground-Tyrant	*Muscisaxicola albilora*	R
☐	Cinnamon-bellied Ground-Tyrant	*Muscisaxicola capistratus*	R
☐	Black-fronted Ground-Tyrant	*Muscisaxicola frontalis*	R
☐	Rufous-webbed Bush-Tyrant	*Polioxolmis rufipennis*	R
☐	Fire-eyed Diucon	*Xolmis pyrope*	R
☐	Black-billed Shrike-Tyrant	*Agriornis montanus*	R
☐	White-tailed Shrike-Tyrant	*Agriornis albicauda*	R
☐	Great Shrike-Tyrant	*Agriornis lividus*	R
☐	Grey-bellied Shrike-Tyrant	*Agriornis micropterus*	R
☐	Chocolate-vented Tyrant	*Neoxolmis rufiventris*	R
☐	Many-coloured Rush-Tyrant	*Tachuris rubrigastra*	R

	HIRUNDINIDAE (SWALLOWS AND MARTINS)		
☐	Blue-and-white Swallow	*Pygochelidon cyanoleuca*	R
☐	Andean Swallow	*Orochelidon andecola*	R
☐	Brown-chested Martin	*Progne tapera*	V
☐	Grey-breasted Martin	*Progne chalybea*	V
☐	Peruvian Martin	*Progne murphyi*	V/M?
☐	Chilean Swallow	*Tachycineta leucopyga*	R
☐	Bank Swallow/Sand Martin	*Riparia riparia*	M
☐	Barn Swallow	*Hirundo rustica*	M
☐	Cliff Swallow	*Petrochelidon pyrrhonota*	M
	TROGLODYTIDAE (WRENS)		
☐	House Wren	*Troglodytes aedon*	R
☐	Grass Wren	*Cistothorus platensis*	R
	TURDIDAE (THRUSHES)		
☐	Austral Thrush	*Turdus falcklandii*	R
☐	Creamy-bellied Thrush	*Turdus amaurochalinus*	V
☐	Chiguanco Thrush	*Turdus chiguanco*	R
☐	Sombre Thrush	*Turdus anthracinus*	R
	MIMIDAE (MOCKINGBIRDS)		
☐	Chilean Mockingbird	*Mimus thenca*	R
☐	Patagonian Mockingbird	*Mimus patagonicus*	R
☐	White-banded Mockingbird	*Mimus triurus*	V
	PASSERIDAE (OLD WORLD SPARROWS)		
☐	House Sparrow	*Passer domesticus*	R
	MOTACILLIDAE (PIPITS)		
☐	Yellowish Pipit	*Anthus lutescens*	V
☐	Correndera Pipit	*Anthus correndera*	R
☐	Hellmayr's Pipit	*Anthus hellmayri*	R
	FRINGILLIDAE (SISKINS)		
☐	Thick-billed Siskin	*Spinus crassirostris*	R
☐	Hooded Siskin	*Spinus magellanicus*	R
☐	Black Siskin	*Spinus atratus*	R
☐	Yellow-rumped Siskin	*Spinus uropygialis*	R
☐	Black-chinned Siskin	*Spinus barbatus*	R
	PASSERELLIDAE (NEW WORLD SPARROWS)		
☐	Rufous-collared Sparrow	*Zonotrichia capensis*	R
	ICTERIDAE (NEW WORLD BLACKBIRDS)		
☐	Bobolink	*Dolichonyx oryzivorus*	V
☐	Peruvian Meadowlark	*Leistes bellicosus*	R

☐	Long-tailed Meadowlark	*Leistes loyca*	R
☐	Screaming Cowbird	*Molothrus rufoaxillaris*	R
☐	Shiny Cowbird	*Molothrus bonariensis*	R
☐	Austral Blackbird	*Curaeus curaeus*	R
☐	Yellow-winged Blackbird	*Agelasticus thilius*	R
	PARULIDAE (NEW WORLD WARBLERS)		
☐	American Redstart	*Setophaga ruticilla*	V
	THRAUPIDAE (TANAGERS, FINCHES AND SEEDEATERS)		
☐	Giant Conebill	*Conirostrum binghami*	R
☐	Tamarugo Conebill	*Conirostrum tamarugense*	R
☐	Cinereous Conebill	*Conirostrum cinereum*	R
☐	Puna Yellow-Finch	*Sicalis lutea*	R
☐	Bright-rumped Yellow-Finch	*Sicalis uropygialis*	R
☐	Greater Yellow-Finch	*Sicalis auriventris*	R
☐	Greenish Yellow-Finch	*Sicalis olivascens*	R
☐	Patagonian Yellow-Finch	*Sicalis lebruni*	R
☐	Saffron Finch	*Sicalis flaveola*	R
☐	Grassland Yellow-Finch	*Sicalis luteola*	R
☐	Raimondi's Yellow-Finch	*Sicalis raimondii*	M/R?
☐	Black-hooded Sierra-Finch	*Phrygilus atriceps*	R
☐	Grey-hooded Sierra-Finch	*Phrygilus gayi*	R
☐	Patagonian Sierra-Finch	*Phrygilus patagonicus*	R
☐	Mourning Sierra-Finch	*Phrygilus fruticeti*	R
☐	Plumbeous Sierra-Finch	*Geospizopsis unicolor*	R
☐	Ash-breasted Sierra-Finch	*Geospizopsis plebejus*	R
☐	Band-tailed Sierra-Finch	*Porphyrospiza alaudina*	R
☐	Red-backed Sierra-Finch	*Idiopsar dorsalis*	R
☐	White-throated Sierra-Finch	*Idiopsar erythronotus*	R
☐	White-winged Diuca-Finch	*Idiopsar speculifera*	R
☐	White-bridled Finch	*Melanodera melanodera*	R
☐	Yellow-bridled Finch	*Melanodera xanthogramma*	R
☐	Band-tailed Seedeater	*Catamenia analis*	R
☐	Black-throated Flowerpiercer	*Diglossa brunneiventris*	R
☐	Blue-black Grassquit	*Volatinia jacarina*	R
☐	Chestnut-throated Seedeater	*Sporophila telasco*	R
☐	Golden-billed Saltator	*Saltator aurantiirostris*	R
☐	Slender-billed Finch	*Xenospingus concolor*	R
☐	Common Diuca-Finch	*Diuca diuca*	R
☐	Blue-and-yellow Tanager	*Pipraeidea bonariensis*	R

Chilean Mockingbird *Mimus thenca*, Farellones Road, Metropolitan Region, Chile (Gonzalo E. González Cifuentes).

Moustached Turca *Pteroptochos megapodius*, Valparaíso Region, Chile (Rodrigo Gazmuri F.).

INDEX

A

Accipiter chilensis 110, 119
Actitis macularius 88
Aeronautes andecolus 134
Agelasticus thilius 166
Agriornis albicauda 158
    lividus 158
    micropterus 158
    montanus 158
Albatross, Antipodean 24
    Atlantic Yellow-nosed 188
    Black-browed 26, 30, 31, 186
    Buller's 26, 30, 31
    Chatham 28, 30, 31
    Grey-headed 26, 30, 31
    Light-mantled 24
    Northern Royal 22, 30, 31
    Pacific 26, 30, 31
    Salvin's 28, 30, 31
    Southern Royal 22, 30, 31
    Wandering 22, 30, 31
    Waved 24
    White-capped 28, 30, 31
Anairetes fernandezianus 150
    flavirostris 150
    parulus 150
    reguloides 150
Anas bahamensis 68
    flavirostris 68
    georgica 68
Ani, Groove-billed 168
Anous albivitta 58, 61
    stolidus 58, 61
Anthus correndera 170, 187
    hellmayri 170
    lutescens 170
Aphrastura masafuerae 152
    spinicauda 152
Aphrodroma brevirostris 38
Aptenodytes forsteri 18, 186
    patagonicus 18
Ardea alba 80
    cocoi 80
Ardenna bulleri 32
    creatopus 32
    gravis 32
    grisea 32
    pacifica 36

Arenaria interpres 86
Asio flammeus 126
Asthenes anthoides 148
    dorbignyi arequipae 148
    modesta 148
    pudibunda 148
    pyrrholeuca 148
Athene cunicularia 126
Attagis gayi 98
    malouinus 98
Avocet, Andean 84

B

Bartramia longicauda 98
Bittern, Least 190
    Stripe-backed 82
Blackbird, Austral 168
    Yellow-winged 166
Bobolink 191
Booby, Blue-footed 48
    Brown 188
    Masked 48
    Peruvian 48, 186
    Red-footed 188
Bubo magellanicus 126
Bubulcus ibis 80
Burhinus superciliaris 98, 186
Bush-Tyrant, Rufous-webbed 158
Buteo albigula 110, 119
    swainsoni 190
    ventralis 110, 119
Butorides striata 82
Buzzard-Eagle, Black-chested 108, 118

C

Calidris alba 90
    bairdii 90
    canutus 90
    fuscicollis 90
    himantopus 88
    mauri 90
    melanotos 88
    minutilla 90
    pusilla 90
    subruficollis 189
    virgata 86
Callipepla californica 98, 187
Campephilus magellanicus 130, 187

Canastero, Arequipa 148
  Austral 148
  Canyon 148
  Cordilleran 148
  Dusky-tailed 148
  Sharp-billed 148
Caracara, Chimango 116, 119
  Mountain 116, 118, 119
  Southern 116, 118
  Striated 116, 119, 187
  White-throated 116, 118, 119
Caracara plancus 116, 118
Catamenia analis 174
Cathartes aura 102, 118
Chaetura pelagica 134
Charadrius alticola 92
  collaris 92
  falklandicus 92
  modestus 94
  nivosus 92, 187
  semipalmatus 92
  vociferus 92
  wilsonia 189
Chat-Tyrant, D'Orbigny's 160
  White-browed 160
Chilia, Crag 146
Chionis albus 52, 186
Chloephaga hybrida 64
  melanopterus 64
  picta 64
  poliocephala 64, 186
  rubidiceps 64
Chordeiles acutipennis 128
Chroicocephalus cirrocephalus 54, 60
  maculipennis 54, 60
  serranus 54, 60
Cinclodes, Black 142
  Buff-winged 142
  Cream-winged 142
  Dark-bellied 142
  Grey-flanked 142
  Seaside 142
  White-winged 142
Cinclodes albiventris 142
  atacamensis 142
  fuscus 142
  maculirostris 142
  nigrofumosus 142
  oustaleti 142
  patagonicus 142
Circus buffoni 190

  cinereus 112, 118, 119
Cistothorus platensis 170
Colaptes pitius 130
  rupicola 130
Colibri coruscans 134
Colorhamphus parvirostris 162
Columba livia 120
Columbina cruziana 122
  picui 122
  talpacoti 122
Condor, Andean 102, 118, 187
Conebill, Cinereous 172
  Giant 172
  Tamarugo 172
Conirostrum binghami 172
  cinereum 172
  tamarugense 172
Coot, Andean 74
  Giant 74
  Horned 74
  Red-fronted 74
  Red-gartered 74, 187
  White-winged 74
Coragyps atratus 102, 118
Cormorant, Antarctic 44
  Guanay 44
  Imperial 44
  Magellanic 44
  Neotropic 44, 186
  Red-legged 44
Coscoroba coscoroba 62
Cowbird, Screaming 168
  Shiny 168, 187
Crake, Dot-winged 190
Creagrus furcatus 52, 60
Crotophaga sulcirostris 168
Curaeus curaeus 168
Curlew, Bristle-thighed 189
Cyanoliseus patagonus 124
Cygnus melancoryphus 62, 186

D

Daption capense 32
Dendrocygna autumnalis 62
  bicolor 62
  viduata 62
Diglossa brunneiventris 172
Diomedea antipodensis 24
  epomophora 22, 30, 31
  exulans 22, 30, 31
  sanfordi 22, 30, 31

Diuca diuca 178, 187
Diuca-Finch, Common 178, 187
    White-winged 176
Diucon, Fire-eyed 162
Diving Petrel, Common 42
    Magellanic 42
    Peruvian 42, 186
Dolichonyx oryzivorus 191
Doradito, Ticking 152
Dotterel, Rufous-chested 94
    Tawny-throated 94
Dove, Eared 122, 187
    Rock 120
    West Peruvian 120
Dowitcher, Short-billed 88
Duck, Andean 70
    Black-headed 70
    Crested 66
    Lake 70
    Spectacled 66
    Torrent 66

E

Earthcreeper, Band-tailed 146
    Buff-breasted 146
    Patagonian Forest 146
    Scale-throated 146
    Straight-billed 146
    White-throated 146
Egret, Cattle 80
    Great 80
    Snowy 80
Egretta caerulea 80
    thula 80
    tricolor 82
Elaenia, White-crested 162, 187
Elaenia albiceps 162, 187
Elanus leucurus 112, 119
Enicognathus ferrugineus 124
    leptorhynchus 124, 187
Eudromia elegans 100
Eudyptes chrysocome 20
    chrysolophus 20
Eudyptula minor 20
Eugralla paradoxa 140
Eulidia yarrellii 134, 187

F

Falco femoralis 114, 119
    peregrinus 114, 119
    sparverius 114, 119

Falcon, Aplomado 114, 119
    Peregrine 114, 119
Field-Tyrant, Short-tailed 191
Finch, Saffron 182
    Slender-billed 172
    White-bridled 180
    Yellow-bridled 180
Firecrown, Green-backed 136, 187
    Juan Fernandez 136
Flamingo, Andean 78
    Chilean 78, 186
    Puna 78
Flicker, Andean 130
    Chilean 130
Flowerpiercer, Black-throated 172
Flycatcher, Fork-tailed 160
    Rufescent 160
    Vermilion 162
Fregata minor 46
Fregetta grallaria 42
    tropica 42
Frigatebird, Great 46
Fulica ardesiaca 74
    armillata 74, 187
    cornuta 74
    gigantea 74
    leucoptera 74
    rufifrons 74
Fulmar, Southern 32
Fulmarus glacialoides 32

G

Gallinago andina 96
    magellanica 96, 187
    stricklandii 96
Gallinula galeata 76
    melanops 76
Gallinule, Common 76
    Purple 76
    Spot-flanked 76
Garrodia nereis 42
Geositta antarctica 144
    cunicularia 144, 187
    isabellina 144
    maritima 144
    punensis 144
    rufipennis 144
Geospizopsis plebejus 176
    unicolor 176
Geranoaetus melanoleucus 108
    polyosoma 104, 118, 119, 187

polyosoma exsul 106
polyosoma poecilochrous 106, 118, 119
Glaucidium nana 128
   peruanum 128
Godwit, Hudsonian 86
   Marbled 86
Goose, Andean 64
   Ashy-headed 64, 186
   Kelp 64
   Ruddy-headed 64
   Upland 64
Grassquit, Blue-black 174
Grebe, Great 72
   Hooded 72
   Northern Silvery 72
   Pied-billed 72, 186
   Southern Silvery 72
   White-tufted 72
Ground-Dove, Bare-faced 122
   Black-winged 120
   Croaking 122
   Golden-spotted 120
   Picui 122
   Ruddy 122
Ground-Tyrant, Black-fronted 156
   Cinereous 156
   Cinnamon-bellied 154
   Dark-faced 154
   Ochre-naped 156
   Puna 156
   Rufous-naped 156
   Spot-billed 154
   White-browed 156
   White-fronted 156
Gull, Andean 54, 60
   Belcher's 52, 60
   Brown-hooded 54, 60
   Dolphin 52, 60
   Franklin's 54, 60
   Grey 52, 60, 186
   Grey-hooded 54, 60
   Kelp 52, 60, 186
   Laughing 54, 60
   Sabine's 54, 60
   Swallow-tailed 52, 60
Gygis alba 58, 61

H

Haematopus ater 84
   leucopodus 84
   palliatus 84, 187

Halobaena caerulea 34, 186
Harrier, Cinereous 112, 118, 119
   Long-winged 190
Hawk, Chilean 110, 119
   Harris's 108, 119
   Masafuera 106
   Puna 106, 118, 119
   Rufous-tailed 110, 119
   Swainson's 190
   Variable 104, 118, 119, 187
   White-throated 110, 119
Heron, Black-crowned Night 80, 186
   Cocoi 80
   Little Blue 80
   Striated 82
   Tricoloured 82
   Yellow-crowned Night 80
Heteronetta atricapilla 70
Hillstar, Andean 136
   White-sided 136
Himantopus mexicanus 84, 187
Hirundo rustica 132
Huet-huet, Black-throated 138
   Chestnut-throated 138
Hummingbird, Giant 136
   Oasis 134
Hymenops perspicillatus 154

I

Ibis, Andean 82
   Black-faced 82, 186
   Puna 82
   White-faced 82
Idiopsar dorsalis 176
   erythronotus 176
   speculifera 176
Ixobrychus exilis 190
   involucris 82

J

Jaeger, Long-tailed 50
   Parasitic 50
   Pomarine 50

K

Kestrel, American 114, 119
Killdeer 92
Kingbird, Eastern 160
   Tropical 160
Kingfisher, Ringed 130, 187
Kiskadee, Great 160

Kite, White-tailed 112, 119
Knot, Red 90

L

Lapwing, Andean 94
    Southern 94, 187
Larosterna inca 58, 61
Larus belcheri 52, 60
    dominicanus 52, 60, 186
Laterallus jamaicensis 76
Leistes bellicosus 166
    loyca 166, 187
Leptasthenura aegithaloides 150
    berlepschi 150
    pallida 150
    striata 150
Lessonia oreas 154
    rufa 154, 187
Leucophaeus atricilla 54, 60
    modestus 52, 60, 186
    pipixcan 54, 60
    scoresbii 52, 60
Limnodromus griseus 88
Limosa fedoa 86
    haemastica 86
Lophonetta specularioides 66

M

Macronectes giganteus 24
    halli 24
Mareca sibilatrix 68
Martin, Brown-chested 191
    Grey-breasted 191
    Peruvian 134
    Sand 132
Meadowlark, Long-tailed 166, 187
    Peruvian 166
Megaceryle torquata 130, 187
Melanodera melanodera 180
    xanthogramma 180

Merganetta armata 66
Metriopelia aymara 120
    ceciliae 122
    melanoptera 120
Milvago chimango 116, 119
Mimus patagonicus 164
    thenca 164, 187
    triurus 164
Miner, Common 144, 187
    Creamy-rumped 144

    Greyish 144
    Puna 144
    Rufus-banded 144
    Short-billed 144
Mockingbird, Chilean 164, 187
    Patagonian 164
    White-banded 164
Molothrus bonariensis 168, 187
    rufoaxillaris 168
Muscigralla brevicauda 191
Muscisaxicola albifrons 156
    albilora 156
    capistratus 154
    cinereus 156
    flavinucha 156
    frontalis 156
    juninensis 156
    maclovianus 154
    maculirostris 154
    rufivertex 156
Myiophobus rufescens 160
Myiopsitta monachus 124

N

Negrito, Andean 154
    Austral 154, 187
Neoxolmis rufiventris 158
Nesofregetta fuliginosa 42
Netta peposaca 70
Nighthawk, Lesser 128
Nightjar, Band-winged 128, 187
    Tschudi's 128
Noddy, Brown 58, 61
    Grey 58, 61
Nothoprocta ornata 100
    perdicaria 100, 187
Numenius phaeopus 86
    tahitiensis 189
Nyctanassa violacea 80
Nycticorax nycticorax 80, 186
Nycticryphes semicollaris 96, 187

O

Oceanites gracilis 40, 186
    oceanicus 40, 186
    pincoyae 40
Oceanodroma hornbyi 40
    markhami 40
    tethys 40
Ochetorhynchus melanurus 146
    phoenicurus 146

ruficaudus 146
Ochthoeca leucophrys 160
    oenanthoides 160
Onychoprion fuscatus 58, 61
    lunatus 58, 61
Oreopholus ruficollis 94
Oreotrochilus estella 136
    leucopleurus 136
Orochelidon andecola 132
Osprey 112, 118
Owl, Barn 126, 187
    Burrowing 126
    Magellanic Horned 126
    Rufous-legged 126
    Short-eared 126
Oxyura ferruginea 70
    vitatta 70
Oystercatcher, American 84, 187
    Blackish 84
    Magellanic 84

P

Pachyptila belcheri 34
    desolata 34
Pagodroma nivea 38
Painted-snipe, South American 96, 187
Pandion haliaetus 112, 118
Parabuteo unicinctus 108, 119
Parakeet, Austral 124
    Monk 124
    Mountain 124
    Red-masked 124
    Slender-billed 124, 187
Pardirallus sanguinolentus 76
Parrot, Burrowing 124
Passer domesticus 178, 187
Patagioenas albipennis 120
    araucana 120
Patagona gigas 136
Pelagodroma marina 40
Pelecanoides garnoti 42, 186
    magellani 42
    urinatrix 42
Pelecanus occidentalis 46
    thagus 46, 186
Pelican, Brown 46
    Peruvian 46, 186
Penguin, Adelie 18
    Chinstrap 18
    Emperor 18, 186
    Gentoo 18

Humboldt 20
King 18
Little 20
Macaroni 20
Magellanic 20
Southern Rockhopper 20
Petrel, Antarctic 38
    Atlantic 38
    Black-winged 188
    Blue 34, 186
    Cape 32
    Cook's 34
    Grey 38
    Henderson 36
    Herald 36
    Juan Fernandez 34
    Kerguelen 38
    Kermadec 36
    Masatierra 34
    Mottled 38
    Murphy's 36
    Northern Giant 24
    Parkinson's 188
    Phoenix 36
    Snow 38
    Southern Giant 24
    Stejneger's 34
    Westland 32
    White-chinned 32
    White-headed 38
Petrochelidon pyrrhonota 132
Phaethon aethereus 48
    lepturus 48
    rubricauda 48, 186
Phalacrocorax atriceps 44
    bougainvilliiorum 44
    bransfieldensis 44
    brasilianus 44, 186
    gaimardi 44
    magellanicus 44
Phalarope, Red 86
    Red-necked 86
    Wilson's 88
Phalaropus fulicarius 86
    lobatus 86
Phalcoboenus albogularis 116, 118, 119
    australis 116, 119, 187
    megalopterus 116, 118, 119
Phasianus colchicus 98
Pheasant, Common 98
Phegornis mitchellii 94, 187

Phleocryptes melanops 152, 187
Phoebastria irrorata 24
Phoebetria palpebrata 24
Phoenicoparrus andinus 78
  jamesi 78
Phoenicopterus chilensis 78, 186
Phrygilus atriceps 180
  fruticeti 178
  gayi 180
  patagonicus 180
Phytotoma rara 166, 187
Pigeon, Chilean 120
  Feral 120
  White-winged 120
Pintail, White-cheeked 68
  Yellow-billed 68
Pipit, Correndera 170, 187
  Hellmayr's 170
  Yellowish 170
Pipraeidea bonariensis 174
Pitangus sulphuratus 160
Plantcutter, Rufous-tailed 166, 187
Platalea ajaja 78
Plegadis chihi 82
  ridgwayi 82
Plover, American Golden 96
  Collared 92
  Grey 96
  Magellanic 94, 187
  Puna 92
  Semipalmated 92
  Snowy 92, 187
  Two-banded 92
  Wilson's 189
Pluvialis dominica 96
  squatarola 96
Pluvianellus socialis 94, 187
Pochard, Rosy-billed 70
Podiceps gallardoi 72
  juninensis 72
  major 72
  occipitalis 72
Podilymbus podiceps 72, 186
Polioxolmis rufipennis 158
Porphyrio martinicus 76
Porphyrospiza alaudina 178, 187
Porzana spiloptera 190
Prion, Antarctic 34
  Slender-billed 34
Procellaria aequinoctialis 32
  cinerea 38

parkinsoni 188
westlandica 32
Progne chalybea 191
  murphyi 134
  tapera 191
Pseudasthenes humicola 148
Pseudocolopteryx citreola 152
Psilopsiagon aurifrons 124
Psittacara erythrogenys 124
Pterodroma alba 36
  atrata 36
  cookii 34
  defilippiana 34
  externa 34
  heraldica 36
  incerta 38
  inexpectata 38
  lessonii 38
  longirostris 34
  neglecta 36
  nigripennis 188
  ultima 36
Pteroptochos castaneus 138
  megapodius 138, 187
  tarnii 138
Puffinus elegans 36
  nativitatis 36
  puffinus 38
Pygarrhichas albogularis 130
Pygmy-Owl, Austral 128
  Peruvian 128
Pygochelidon cyanoleuca 132
Pygoscelis adeliae 18
  antarcticus 18
  papua 18
Pyrocephalus rubinus 162

Q

Quail, California 98, 187

R

Rail, Austral 76
  Black 76
  Plumbeous 76
Rallus antarcticus 76
Rayadito, Masafuera 152
  Thorn-tailed 152
Recurvirostra andina 84
Redstart, American 172
Rhea, Lesser 100, 186
  Puna 100

Rhea pennata 100, 186
   tarapacensis 100
Rhodopis vesper 134
Riparia riparia 132
Rollandia rolland 72
Rushbird, Wren-like 152, 187
Rush-Tyrant, Many-coloured 152, 187
Rynchops niger 56, 61

S

Saltator, Golden-billed 174
Saltator aurantiirostris 174
Sanderling 90
Sandpiper, Baird's 90
   Buff-breasted 189
   Least 90
   Pectoral 88
   Semipalmated 90
   Solitary 88
   Spotted 88
   Stilt 88
   Upland 98
   Western 90
   White-rumped 90
Sandpiper-Plover, Diademed 94, 187
Scelorchilus albicollis 138
   rubecula 140
Scytalopus fuscus 140
   magellanicus 140
Seedeater, Band-tailed 174
   Chestnut-throated 174
Seedsnipe, Grey-breasted 98, 187
   Least 98
   Rufous-bellied 98
   White-bellied 98
Sephanoides fernandensis 136
   sephaniodes 136, 187
Setophaga ruticilla 172
Sheartail, Peruvian 134
Shearwater, Buller's 32
   Christmas 36
   Great 32
   Manx 38
   Pink-footed 32
   Sooty 32
   Subantarctic 36
   Wedge-tailed 36
Sheathbill, Snowy 52, 186
Shoveler, Red 68, 186
Shrike-Tyrant, Black-billed 158
   Great 158

   Grey-bellied 158
   White-tailed 158
Sicalis auriventris 182
   flaveola 182
   lebruni 182
   lutea 182
   luteola 182, 187
   olivascens 182
   raimondii 191
   uropygialis 182
Sierra-Finch, Ash-breasted 176
   Band-tailed 178, 187
   Black-hooded 180
   Grey-hooded 180
   Mourning 178
   Patagonian 180
   Plumbeous 176
   Red-backed 176
   White-throated 176
Siskin, Black 184
   Black-chinned 184, 187
   Hooded 184
   Thick-billed 184
   Yellow-rumped 184
Skimmer, Black 56, 61
Skua, Brown 50
   Chilean 50, 186
   South Polar 50
Snipe, Fuegian 96
   Magellanic 96, 187
   Puna 96
Sparrow, House 178, 187
   Rufous-collared 178, 187
Spatula cyanoptera 70, 186
   discors 190
   platalea 68, 186
   puna 70
   versicolor 70
Speculanas specularis 66
Spheniscus humboldti 20
   magellanicus 20
Spinus atratus 184
   barbatus 184, 187
   crassirostris 184
   magellanicus 184
   uropygialis 184
Spoonbill, Roseate 78
Sporophila telasco 174
Steamerduck, Flying 66
   Magellanic 66
Steganopus tricolor 88

Stercorarius antarcticus 50
   chilensis 50, 186
   longicaudus 50
   maccormicki 50
   parasiticus 50
   pomarinus 50
Sterna hirundinacea 56, 61
   hirundo 56, 61
   paradisaea 56, 61
   trudeaui 58, 61
   vittata 56, 61
Sternula antillarum 189
   lorata 56, 61, 186
Stilt, Black-necked 84, 187
Storm-petrel, Black-bellied 42
   Grey-backed 42
   Markham's 40
   Pincoya 40
   Polynesian 42
   Ringed 40
   Wedge-rumped 40
   White-bellied 42
   White-faced 40
   White-vented 40, 186
   Wilson's 40, 186
Strix rufipes 126
Sula dactylatra 48
   leucogaster 188
   nebouxii 48
   sula 188
   variegata 48, 186
Surfbird 86
Swallow, Andean 132
   Bank 132
   Barn 132
   Blue-and-white 132
   Chilean 132, 187
   Cliff 132
Swan, Black-necked 62, 186
   Coscoroba 62
Swift, Andean 134
   Chimney 134
Sylviorthorhynchus desmursii 152
Systellura decussata 128
   longirostris 128, 187

T

Tachuris rubrigastra 152, 187
Tachycineta leucopyga 132, 187
Tachyeres patachonicus 66
   pteneres 66

Tanager, Blue-and-yellow 174
Tapaculo, Chucao 140
   Dusky 140
   Magellanic 140
   Ochre-flanked 140
   White-throated 138
Tattler, Wandering 189
Teal, Blue-winged 190
   Cinnamon 70, 186
   Puna 70
   Silver 70
   Yellow-billed 68
Tern, Antarctic 56, 61
   Arctic 56, 61
   Common 56, 61
   Elegant 58, 61
   Grey-backed 58, 61
   Inca 58, 61
   Least 189
   Peruvian 56, 61, 186
   Sandwich 58, 61
   Snowy-crowned 58, 61
   Sooty 58, 61
   South American 56, 61
   White 58, 61
Thalassarche bulleri bulleri 26, 30, 31
   bulleri platei 26, 30, 31
   chlororhynchos 188
   chrysostoma 26, 30, 31
   eremita 28, 30, 31
   melanophris 26, 30, 31, 186
   salvini 28, 30, 31
   steadi 28, 30, 31
Thalasseus elegans 58, 61
   sandvicensis 58, 61
Thalassoica antarctica 38
Thaumastura cora 134
Theristicus branickii 82
   melanopis 82, 186
Thick-knee, Peruvian 98, 186
Thinocorus orbignyianus 98, 187
   rumicivorus 98
Thrush, Austral 164, 187
   Chiguanco 164
   Creamy-bellied 164
   Sombre 164
Tinamotis ingoufi 100
   pentlandii 100, 187
Tinamou, Chilean 100, 187
   Elegant Crested 100
   Ornate 100

Patagonian 100
Puna 100, 187
Tit-Spinetail, Patagonian 150
    Plain-mantled 150
    Puna 150
    Streaked 150
Tit-Tyrant, Juan Fernandez 150
    Pied-crested 150
    Tufted 150
    Yellow-billed 150
Treerunner, White-throated 130
Tringa flavipes 88
    incana 189
    melanoleuca 88
    semipalmata 86
    solitaria 88
Troglodytes aedon 170, 187
Tropicbird, Red-billed 48
    Red-tailed 48, 186
    White-tailed 48
Turca, Moustached 138, 187
Turdus amaurochalinus 164
    anthracinus 164
    chiguanco 164
    falcklandii 164, 187
Turnstone, Ruddy 86
Tyrannus melancholicus 160
    savana 160
    tyrannus 160
Tyrant, Chocolate-vented 158
    Patagonian 162
    Spectacled 154
Tyto alba 126, 187

U

Upucerthia albigula 146
    dumetaria 146
    saturatior 146
    validirostris 146

V

Vanellus chilensis 94, 187
    resplendens 94

Veniliornis lignarius 130
Violetear, Sparkling 134
Volatinia jacarina 174
Vultur gryphus 102, 118, 187
Vulture, Black 102, 118
    Turkey 102, 118

W

Whimbrel 86
Whistling Duck, Black-bellied 62
    Fulvous 62
    White-faced 62
Wigeon, Chiloe 68
Willet 86
Wiretail, Des Murs's 152
Woodpecker, Magellanic 130, 187
    Striped 130
Woodstar, Chilean 134, 187
Wren, Grass 170
    House 170, 187

X

Xema sabini 54, 60
Xenospingus concolor 172
Xolmis pyrope 162

Y

Yellow-Finch, Bright-rumped 182
    Grassland 182, 187
    Greater 182
    Greenish 182
    Patagonian 182
    Puna 182
    Raimondi's 191
Yellowlegs, Greater 88
    Lesser 88

Z

Zenaida auriculata 122, 187
    meloda 120
Zonotrichia capensis 178, 187

Chilean Tinamou *Nothoprocta perdicaria*, Cuesta Buenos Aires, Coquimbo Region, Chile (Jose F. Cañas).

QUICK INDEX TO THE MAIN GROUPS OF BIRDS

Figures in **bold** refer to plate numbers

Albatrosses	3–8	Harriers	50	Rheas	44		
Anis	79	Hawks	46–49	Rush-Tyrant	71		
Avocets	36	Herons	34–35	Rushbird	71		
Bitterns	35	Hillstars	63	Saltators	82		
Blackbirds	78–79	Huet-huets	64	Sandpipers	37–39, 43		
Boobies	17	Hummingbirds	62–63	Seedeaters	82		
Bridled Finches	85	Ibises	35	Seedsnipes	43		
Buzzard-Eagle	48	Jaegers	18	Shearwaters	9, 11, 12		
Canasteros	69	Kestrels	51	Sheathbills	19		
Caracaras	52	Kingbirds	75	Shrike-Tyrants	74		
Chat-Tyrants	75	Kingfishers	60	Sierra-Finches	83–85		
Cinclodes	66	Kites	50	Siskins	87		
Condor	45	Lapwings	41	Skimmers	21, 24		
Conebills	81	Martins	62	Skuas	18		
Coots	31	Meadowlarks	78	Snipes	42		
Cormorants	15	Miners	67	Sparrows	84		
Cowbirds	79	Mockingbirds	77	Spoonbills	33		
Diuca-Finches	83–84	Negritos	72	Steamerducks	27		
Diving Petrels	14	Night Herons	34	Stilts	36		
Doraditos	71	Nightjars	59	Storm-Petrels	13–14		
Dotterels	41	Noddies	22, 24	Swallows	61		
Doves	55–56	Osprey	50	Swans	25		
Ducks	27–29	Ovenbirds	66–71	Swifts	62		
Earthcreepers	68	Owls	58–59	Tanagers	82		
Egrets	34	Oystercatchers	36	Tapaculos	64–65		
Elaenias	76	Painted-snipe	42	Terns	21–22, 24		
Falcons	51	Parakeets	57	Thick-knees	43		
Finches	81, 85	Parrots	57	Thrushes	77		
Firecrowns	63	Pelicans	16	Tinamous	44		
Flamingos	33	Penguins	1–2	Tit-Spinetails	70		
Flickers	60	Petrels	9–12	Tit-Tyrants	70		
Flowerpiercers	81	Phalaropes	37–38	Treerunner	60		
Flycatchers	75–76	Pheasants	43	Tropicbirds	17		
Frigatebirds	16	Pigeons	55	Tyrant-Flycatchers	70–76		
Gallinules	32	Pipits	80	Vultures	45		
Geese	26	Plantcutters	78	Whistling Ducks	25		
Giant Petrels	4	Plovers	40–42	Wiretail	71		
Godwits	37	Prions	10	Woodpeckers	60		
Grassquits	82	Pygmy-Owls	59	Wrens	80		
Grebes	30	Quails	43	Yellow-Finches	86		
Ground-Doves	55–56	Rails	32	Yellowlegs	38		
Ground-Tyrants	72–73	Rayaditos	71				
Gulls	19–20, 23	Redstarts	81				